Lecture Notes in Mathematics

A collection of informal reports and seminars
Edited by A. Dold, Heidelberg and B. Eckmann, Zürich

123

T0220019

A. V. Jategaonkar

Dept. of Mathematics, Cornell University,
Ithaca, NY/USA

Left Principal Ideal Rings

Springer-Verlag
Berlin · Heidelberg · New York 1970

PREFACE

The aim of this monograph is to present a self-contained account of the structure theory and the ideal theory of principal left ideal rings with unity (pli-rings, for short). It is not an account of all aspects of pli-rings. Indeed, it contains nothing at all concerning modules over pli-rings and very little concerning factorization of elements.

A glance at the table of contents will reveal a little of what the monograph contains. Moreover, each chapter has a separate introduction. In each section, the main results are stated as soon as enough definitions are given to make the statements meaningful. This arrangement may help to get a detailed idea of the contents unhampered by proofs.

A first year graduate course in algebra covering Artinian rings and an acquaintance with ordinals would be enough to read most of the monograph.

Some parts of this monograph are taken from my thesis written under the supervision of Professor Newcomb Greenleaf. I wish to thank him for the encouragement and advice I received from him. Thanks are due to Professor R. E. Johnson who read the first draft of this monograph and made a number of useful suggestions. In particular, in §6 of Chapter II, I have followed the simple and elegant method suggested by him. I wish to thank Professor Alex Rosenberg and Professor P. J. Hilton for their interest in this monograph. I wish to thank Mrs. Manju Bewtra for pointing out some inaccuracies in the first draft.

CONTENTS

CHAPTER I.

LEFT GOLDIE RINGS.

INTRODUCTION.

The aim of this chapter is to lay down a foundation for the study of principal left ideals rings, which will be taken up from the second chapter. The main results of this chapter are Goldie's (first and) second theorem and Small's theorem. If the reader is familiar with these theorems, he should pass over this chapter.

We have proved the main results of this section in their full generality although we need only special cases of these later on. However, we have not included many interesting results concerning left Goldie rings on the (shaky) ground that we do not need them. In other words, there is a lot more to left Goldie rings than is included in here; A few references to the literature are given in the notes at the end of §§2 and 4.

§0. TERMINOLOGY AND NOTATION.

In this section, we state some basic definitions, explain some notational conventions and recall some well-known results.

All rings are assumed to be associative. As usual, unity of a ring (if it has one) is assumed to be non-zero. Except for chapter 1, we shall deal with unitary rings and unitary modules only. Ideal shall mean a two-sided ideal. All one-sided definitions and results are stated from the left; the right analogues are used without any further mention.

If A is a non-empty subset of a ring R, the left annihilator of A in R is defined to be the set

$$\ell_R(A) = \{r \in R \mid ra = 0 \text{ for every } a \in A\} .$$

We shall use $\ell(A)$ instead of $\ell_R(A)$ if there is no confusion; if $A = \{a\}$, we shall use $\ell(a)$ instead of $\ell(\{a\})$. The right annihilator of A in R is denoted by $r_R(A)$ or $r(A)$. It is clear that $\ell_R(A)$ is a left ideal of R; if A is a left ideal of R then $\ell_R(A)$ is an ideal of R. A left ideal ℓ of R is called an annihilator left ideal of R if $\ell = \ell_R(A)$ for some non-empty subset A of R. An annihilator ideal of R is an ideal of R which is also an annihilator left ideal of R.

An element $c \in R$ is called a left regular element in R if $\ell_R(c) = (0)$; c is right regular if $r_R(c) = (0)$; c is regular if $\ell_R(c) = r_R(c) = (0)$.

A left ideal ℓ is nilpotent if $\ell^n = (0)$ for some $n \in Z^+$; the least such n is called the index of nilpotency of ℓ.

An ideal P of a ring R is called a prime ideal if, for $a, b \in R$, $aRb \subseteq P$ implies either $a \in P$ or $b \in P$; equivalently, P is a prime ideal of R if, for two-sided ideals A, B of R, $AB \subseteq P$ implies either $A \subseteq P$ or $B \subseteq P$. A ring R is called a prime ring if

(0) is a prime ideal of R . It is clear that R is a prime ring if and only if $\ell_R(A) = (0)$ for every non-zero left ideal A of R . A domain is a non-zero ring without non-zero zero-divisors. A domain is evidently a prime ring.

For an arbitrary ring R , the intersection of all prime ideals of R is called the prime radical of R and is denoted by P(R) . It is well-known that P(R) is a nil ideal, every nilpotent one-sided ideal of R is contained in P(R) and P(R/P(R)) = (0) . A ring R is called a semi-prime ring if P(R) = (0) . It is well-known that a ring R is semi-prime if and only if R has no non-zero nilpotent ideals. McCoy [1] contains a very readable account of prime radical and related things.

A ring R is called a semi-pli-ring (resp. ipli-ring) if R has unity and if every finitely generated left ideal (resp. every two-sided ideal) of R can be expressed in the form Ra for some $a \in R$. A ring R is called a pli-ring (short for principal left ideal ring) if R has unity and every left ideal of R can be expressed as Ra for some $a \in R$. A ring R is called a left Noetherian ring if R satisfies the ascending chain condition on left ideals. It is well-known that R is left Noetherian if and only if R satisfies the maximum condition on left ideals if and only if every left ideal of R is finitely generated. A ring R is called a left Artinian ring if R has unity and satisfies the descending chain condition (or equivalently, the minimum condition) on left ideals. It is well-known that a left Artinian ring is a left Noetherian ring. cf. Lambek [1], page 69. A semi-prime (resp. prime) left Artinian ring is called a semi-simple (resp. simple) Artinian ring. The well-known Wedderburn-Artin theorem shows that a semi-simple Artinian ring is isomorphic with a direct sum of a finite number of full matrix rings over skew fields and that everything involved is essentially unique.

A ring R is called a left Goldie ring if R satisfies the

ascending chain condition annihilator left ideals and has no infinite direct sums of non-zero left ideals. Commutative integral domains and left Noetherian rings are important examples of left Goldie rings. Observe that the classes of semi-pli-rings, ipli-rings, pli-ring, left Noetherian rings and left Artinian rings are all closed under formation of finite direct sums and homomorphic images. Although a finite direct sum of left Goldie rings with unity is again left Goldie, a homomorphic image of left Goldie ring need not be a left Goldie ring. This prompts us to make the following definition: If every homomorphic image of a ring R is a left Goldie ring then R is called a fully left Goldie ring.

Let A, B be sets. $A\setminus B$ denotes the set of all those elements of A which are not in B. $A \subseteq B$ means A is a (possibly improper) subset of B, whereas $A \subset B$ means $A \subseteq B$ and $A \neq B$; sometimes, for the sake of emphasis, we write $A \subsetneq B$ instead of $A \subset B$. Z (resp. Z^+) denotes the set of all (resp. all strictly positive) integers. ω is the first infinite ordinal number.

§1. LEFT QUOTIENT RINGS.

The first attempt to define rings of quotients of non-commutative rings was made by Ore [1]. The definition given below is essentially due to him.

A subset S of a ring R is called a semi-group of regular elements in R if S is a non-empty multiplicatively closed subset of R and every element of S is regular in R . A <u>left divisor set</u> in R is a semi-group of regular elements in R such that, for every $(s,r) \in S \times R$, there exists $(s_1,r_1) \in S \times R$ with

$$s_1 r = r_1 s \ .$$

If a ring R has at least one regular element and if the semi-group of all regular elements of R is a left divisor set in R then R is said to have the <u>left common multiple property</u>. Observe that every commutative ring containing at least one regular element has the left common multiple property whereas a free algebra on more than one letters over a field has neither the left nor the right common multiple property.

Let S be a semi-group of regular elements in a ring R . A ring Q is called a <u>classical left quotient ring of R w.r.t. S</u> (abbreviated as '<u>l.q. ring of R w.r.t. S</u>') if Q has unity, R is a subring of Q , every element of S is a unit in Q and every $q \in Q$ can be expressed in at least one way as $q = s^{-1}r$, $(s,r) \in S \times R$. If R has at least one regular element and if S is the semi-group of all regular elements of R then a l.q. ring of R w.r.t. S (if such a thing exists) is called a <u>total l.q. ring of R</u> . If Q is a total l.q. ring of R then R is called a <u>left order in Q</u> .

Observe that if S is a semi-group of regular elements in a ring R and if R has a l.q. ring w.r.t. S then S is necessarily a left divisor set in R . It follows that a free algebra on more than one letters over a field cannot have a total l.q. ring.

The main result of this section shows that a ring R has a l.q. ring w.r.t. a semi-group S of regular elements in R if and only if S is a left divisor set in R .

We firstly prove the following rather technical result; we shall need it frequently later on so we give it a name.

<u>1·1. COMMON DENOMINATOR THEOREM</u>. Let S be a semi-group of regular elements in a ring R and let Q be a l.q. ring of R w.r.t.S. Let $s_1, \ldots, s_n \in S$. Then there exist $r_1, \ldots, r_n \in R$ and $s \in S$ such that

$$s_i^{-1} = s^{-1} r_i$$

for $1 \leq i \leq n$.

<u>Proof</u>. For $n = 1$, take $s = s_1^2$ and $r_1 = s_1$. Suppose $r_1^*, \ldots, r_{n-1}^* \in R$ and $s^* \in S$ are obtained such that

$$s_i^{-1} = s^{*-1} r_i^*$$

for $1 \leq i \leq n-1$. It is easily seen that S is a left divisor set in R ; so, there exists $(r_n, r) \in S \times R$ such that $r_n s_n = rs^* = s$, say. Clearly, $s \in S$ and r is a unit in Q . Also, $s_n^{-1} = s^{-1} r_n$. Set $r_i = rr_i^*$ for $1 \leq i \leq n-1$ so that

$$s_i^{-1} = s^{*-1} r_i^* = (rs^*)^{-1}(rr_i^*) = s^{-1} r_i$$

for $1 \leq i \leq n-1$. This completes the induction and concludes the proof.

Notice the following use of the theorem. If Q_1 and Q_2 are l.q. rings of R w.r.t. S then there exists an isomorphism $\sigma\colon Q_1 \to Q_2$ such that $\sigma|R = \mathrm{id}._R$; the map σ is the obvious one and is unique; the common denominator theorem is required to show that σ is a homomorphism of additive groups of Q_1 and Q_2 .

It is convenient to prove the following technicality at this point.

1·2. LEMMA. Let S be a left divisor set in a ring R . Suppose $(c,x),(d,y)$ and (s,r) are elements of $S \times R$ such that

$$rc = sd \quad \text{and} \quad rx = sy .$$

Then $ac = bd$ implies $ax = by$ for all $a,b\in R$.

Proof. Since S is a left divisor set in R , there exists $(s_1,r_1)\in S\times R$ such that $s_1 b = r_1 s$. Thus, $s_1 ac = s_1 bd = r_1 sd = r_1 rc$. Since c is regular in R , we have $s_1 a = r_1 r$; so, $s_1 ax = r_1 rx = r_1 sy = s_1 by$. Since s_1 is regular in R , we get $ax = by$. This completes the proof.

We are now in a position to prove the main result of this section.

1·3. ORE'S THEOREM. Let S be a semi-group of regular elements in a ring R . A l.q. ring of R w.r.t. S exists if and only if S is a left divisor set in R . If a l.q. ring of R w.r.t. S exists then it is essentially unique.

Proof. We have already seen that if a l.q. ring Q of R w.r.t. S exists then it is essentially unique and that S is a left divisor set in R .

Suppose S is a left divisor set in R . We shall construct a l.q. ring Q of R w.r.t. S . The construction we give is a modification of the well-known construction of the field of quotients of a commutative integral domain.

Firstly, we define an equivalence relation on the set $S \times R$.

Let (c,x) and (d,y) be elements of $S \times R$. Then $(c,x) \sim (d,y)$ if and only if, for every $(s,r) \in S \times R$ with $rc = sd$, we have $rx = sy$. Thus, by lemma 1·2, $(c,x) \sim (d,y)$ if and only if there exists $(s',r') \in S \times R$ such that $r'c = s'd$ and $r'x = s'y$. Also observe that since S is a left divisor set in R, we can always find $(s',r') \in S \times R$ with $r'c = s'd$ for any $c,d \in S$. In the following construction, these observations will be used without explicit mention.

We now show that $\sim$ is an equivalence relation on $S \times R$. $\sim$ is clearly reflexive. Suppose $(c,x) \sim (d,y)$. Let $(s_1,r_1) \in S \times R$ with $r_1 d = s_1 c$. Then, by lemma 1·2, we have $r_1 y = s_1 x$; so, $(d,y) \sim (c,x)$. Suppose $(c,x) \sim (d,y)$ and $(d,y) \sim (e,z)$. Pick $(s,r) \in S \times R$ with $rc = se$. There exists $(s_2,r_2) \in S \times R$ such that $r_2 d = s_2 se = s_2 rc$. By 1·2, this implies $r_2 y = s_2 rx$. Similarly, from $(d,y) \sim (e,z)$ and $r_2 d = s_2 se$, we obtain $r_2 y = s_2 sz$. Thus, $s_2 rx = s_2 sz$. Since s_2 is regular in R, we have $rx = sz$. Hence, $(c,x) \sim (e,z)$.

Denote the equivalence class of (c,x) under $\sim$ by x/c. Let

$$Q = \{x/c \mid (c,x) \in S \times R\} .$$

We shall define operations in Q so as to make it a l.q. ring of R w.r.t. S.

Given x/c and y/d in Q, we define

$$(x/c) + (y/d) = (rx + sy)/sd ,$$

where $(s,r) \in S \times R$ with $rc = sd$. We have to show that $+$ is well-defined. Let $(s_1,r_1) \in S \times R$ with $r_1 c = s_1 d$. Then

$$(sd, rx + sy) \sim (s_1 d, r_1 x + s_1 y) .$$

For, let $(s_2,r_2) \in S \times R$ with $r_2 sd = s_2 s_1 d$; regularity of d gives

$r_2s = s_2s_1$; since $rc = sd$ and $r_1c = s_1d$, we have $r_2rc = s_2r_1c$; so, $r_2r = s_2r_1$. Now,

$$r_2(rx + sy) = s_2r_1x + s_2s_1y = s_2(r_1x + s_1y) .$$

This shows that $(x/c) + (y/d)$ is independent of the choice of (s,r). Now, let $(c,x) \sim (c',x')$ and $(d,y) \sim (d',y')$. Pick $(s',r') \in S \times R$ such that $r'c' = s'd'$. We then have

$$(sd, rx + sy) \sim (s'd', r'x' + s'y') .$$

For, choose $(s_3,r_3) \in S \times R$ with $r_3sd = s_3s'd'$; then, by the lemma 1·2 $r_3sy = s_3s'y'$; but, $r_3sd = s_3s'd'$ gives $r_3rc = s_3r'c'$; so, $r_3rx = s_3r'x'$. It follows that

$$r_3(rx + sy) = s_3(r'x' + s'y') .$$

We have thus shown that $+$ is well-defined in Q . It is now straightforward to check that $(Q,+)$ is an abelian group.

Given x/c and y/d in Q , we define

$$(x/c) \cdot (y/d) = (x_1y)/(d_1c) ,$$

where $(d_1,x_1) \in S \times R$ with $d_1x = x_1d$. To see that the multiplication is independent of the choice of (d_1,x_1) , let $(d_2,x_2) \in S \times R$ with $d_2x = x_2d$. Pick $(s,r) \in S \times R$ with $rd_1c = sd_2c$; then $rd_1=sd_2$; so, $rd_1x = sd_2x$ which gives $rx_1d = sx_2d$; so, $rx_1 = sx_2$ and $rx_1y = sx_2y$. Thus, $(d_1c,x_1y) \sim (d_2c,x_2y)$. To see that the multiplication is independent of the representatives of x/c and y/d , let $(c,x) \sim (c',x')$ and $(d,x) \sim (d',x')$. Choose $(d_1',x_1') \in S \times R$ with $d_1'x'=x_1'd'$. We have to show that

$$(d_1c,x_1y) \sim (d_1'c',x_1'y') .$$

Pick $(s_1, r_1) \in S \times R$ with $r_1 d_1 c = s_1 d_1' c'$. By 1·2, $r_1 d_1 x = s_1 d_1' x'$. So, $r_1 x_1 d = s_1 x_1' d'$. Using 1·2 again, we have $r_1 xy = s_1 x_1' y'$. We have thus shown that the multiplication is well-defined in Q . It is now straightforward to check that Q is a ring with unity $1 = c/c$, $c \in S$.

Since $(cx)/c = (dx)/d$ for every $x \in R$ and $c,d \in S$, it follows that

$$r \mapsto (cr)/c \quad , \quad c \in S$$

is a well-defined map from R into Q . This map is easily seen to be a monomorphism. It is now easy to complete the proof.

We note a corollary of some interest. A definition first. A domain D is called a <u>left Ore domain</u> if D has the left common multiple property. Clearly, a domain D is left Ore if and only if $Da \cap Db \neq (0)$ for all non-zero $a,b \in D$.

<u>1·4. COROLLARY</u>. A ring R is a left Ore domain if and only if it is a left order in a skew field.

Since skew fields are well-behaved among non-commutative rings, the interest in left Ore domains is manifest. Which familiar domains are left Ore? Evidently, every commutative domain is left Ore. To single out another familiar class of left Ore domains, we need the following characterization.

<u>1·5. PROPOSITION</u>. A ring R is a left Ore domain if and only if R is a left Goldie domain.

<u>Proof</u>. Let R be a domain which is not left Ore. Then there exist non-zero elements a,b in R such that $Ra \cap Rb = (0)$. It is easily seen that $\sum_{n>0} Rab^n$ is a direct sum of non-zero left ideals of R. Thus, a left Goldie domain is left Ore. A left Ore domain is triv-

ially left Goldie. This completes the proof.

1·6. COROLLARY. Every left Noetherian domain (in particular, every pli-domain) is left Ore.

References.

Ore [1] was the first to realize the usefulness of the left common multiple property; he proved theorem 1·3 in case of domains. Asano [1] began generalizing Ore's construction. By now, there are quite a few notions of rings of quotients available. Some of these may be found in Faith [1] and Lambek [1] .

Our proof of Ore's theorem 1·3 is based on an idea attributed to A.W. Goldie in Smith [1]. Proposition 1·5 is in Cohn [2].

For an interesting connection between left Ore domains and free algebras, see Jategaonkar [10].

We have not tried to trace the authorship of Ore's theorem. To call it Ore's theorem is certainly inaccurate (but then did Euclid ever dream of Euclidean rings?)

§2. GOLDIE'S THEOREM.

In this section, we shall prove the following two theorems due to A.W. Goldie.

2·1. GOLDIE'S FIRST THEOREM. A ring R is a left order in a simple Artinian ring if and only if R is a non-zero prime left Goldie ring.

2·2. GOLDIE'S SECOND THEOREM. A ring R is a left order in a semi-simple Artinian ring if and only if R is a non-zero semi-prime left Goldie ring.

These two theorems together with a theorem due to Faith and Utumi and a theorem of Levitski (cf. §3) form the foundation of the theory of left Noetherian rings.

The 'only if' parts of both theorems of Goldie are relatively easy and we shall dispose them off first.

2·3. LEMMA. If R is a left order in a semi-simple (resp. simple) Artinian ring then R is a non-zero, semi-prime (resp. prime), left Goldie ring.

Proof. Firstly, let us assume that R is a left order in a semi-simple Artinian ring Q. R must be a non-zero ring since a zero-ring has no regular elements and thus cannot have a l.q. ring. If A is a non-empty subset of R then, evidently,

$$\ell_R(A) = \ell_Q(A) \cap R .$$

Since Q has the ascending chain condition on left ideals, R has the ascending chain condition on annihilator left ideals. If $\sum_{n\geq 1} A_n$ is an infinite direct sum of non-zero left ideals of R then, using the common denominator theorem, it follows that $\sum_{n\geq 1} QA_n$ is an infinite direct sum of non-zero left ideals of Q ; this is clearly impossible. Hence R is a non-zero left Goldie ring.

Suppose N is a non-zero nilpotent two-sided ideal of R. Let $k+1$ be the index of nilpotency of N ; so, $k > 0$. Since QNQ is an ideal of Q, there exists a (non-zero) central idempotent e in Q such that $QNQ = Qe$. Using the common denominator theorem, we can express e as

$$e = c^{-1} (\Sigma r_i z_i q_i) ,$$

where $c, r_i \in R$, c regular in R, $z_i \in N$ and $q_i \in Q$. Thus, $ce = ec \in NQ$; consequently, $N^k ec \subseteq N^{k+1} Q = (0)$. Since c is regular in R, it is a unit in Q. It follows that $N^k e = (0) = eN^k$. Now,

$$(QN^kQ)^2 \subseteq QNQN^kQ \subseteq QeN^kQ = (0) .$$

Since Q is semi-simple, we have $QN^kQ = (0)$ so that $N^k = (0)$, contrary to our choice of k. Hence R is semi-prime.

Now suppose that R is a left order in a simple Artinian ring Q. Since Q is also semi-simple, the argument above shows that R is a non-zero left Goldie ring. It remains to show that R is a prime ring. Let A and B be two-sided ideals of R such that $AB = (0)$ but $B \neq (0)$. Since Q is simple, $QBQ = Q$. Using the common denominator theorem, we have

$$1 = c^{-1} (\Sigma b_i q_i) ,$$

i.e.,

$$c = \Sigma b_i q_i ,$$

where $b_i \in B$, $q_i \in Q$ and c is a regular element in R. Now, $(0) = AB = ABQ$ implies $(0) = Ac$ which in turn implies $A = (0)$. Hence R is a prime ring. This completes the proof.

We now proceed to prove the non-trivial 'if' part of Goldie's second theorem. The proof we give is due to Procesi and Small [1]. The main step in the proof is to show that a non-zero semi-prime left Goldie ring has regular elements and has the left common multiple property. Once this is proved, we invoke Ore's theorem to get a total l.q. ring Q; it is then easy to show that Q has to be a semi-simple Artinian ring. The 'if' part of Goldie's first theorem is easily obtained by using the 'if' part of Goldie's second theorem.

2·4. LEMMA. (i) Let A and B be non-empty subsets of a ring R. Then $\ell r \ell(A) = \ell(A)$; $r \ell r(A) = r(A)$; $\ell(A \cup B) = \ell(A) \cap \ell(B)$; further, if $A \subseteq B$, then $\ell(A) \supseteq \ell(B)$.

(ii) R has the ascending chain condition on annihilator left ideals if and only if R has the descending chain condition on annihilator right ideals.

Proof. (i) is straightforward and (ii) follows from it.

2·5. LEMMA. Let R be a ring with the ascending chain condition on annihilator left ideals. For every $a \in R$, there exists $t \in Z^+$ such that

$$\ell(a^n) \cap Ra^m = (0)$$

for every $n \in Z^+$ and every $m \geq t$.

Proof. Choose $t \in Z^+$ such that $\ell(a^t) = \ell(a^m)$ for every $m \geq t$. If $x \in \ell(a^t) \cap Ra^t$, then $x = ya^t$ for some $y \in R$ and $0 = xa^t = ya^{2t}$; thus, $y \in \ell(a^{2t}) = \ell(a^t)$ and $x = ya^t = 0$. Since $Ra^m \subseteq Ra^t$ for all $m \geq t$ and $\ell(a^n) \subseteq \ell(a^t)$ for all $n \in Z^+$, the lemma follows.

The following lemma is the main tool needed in the proof.

2·6. LEMMA. Let R be a non-zero semi-prime ring with the ascending chain condition on annihilator left ideals. If $A \supseteq B$ are left ideals of R and if $r(A) \subsetneqq r(B)$ then there exists an element $a' \in A$ such that $Aa' \cap B = (0)$ and $Aa' \neq (0)$.

Proof. Let U be an annihilator right ideal minimal with respect to being contained in $r(B)$ and properly containing $r(A)$. By its choice, $AU \neq (0)$; so, $AUAU \neq (0)$ since R is semi-prime. Pick $ua \in UA$ such that $AuaU \neq (0)$. We claim that $Aua \cap B = (0)$. If not, pick $x \in A$ such that $0 \neq xua \in Aua \cap B$. Since $xua \in B$ and $U \subseteq r(B)$, we have $xuaU = (0)$. Since $x \in A, r(x) \supseteq r(A)$. Consider $r(x) \cap U$. It is an annihilator right ideal containing $r(A)$ and contained in U. Since $uaU \subseteq r(x)$, but $uaU \nsubseteq r(A)$, we deduce from the minimality of U that $r(x) \cap U = U$. This implies $xU = (0)$, contrary to $xua \neq 0$. The lemma follows by taking $a' = ua$.

We need a definition. A left ideal A of a ring R is an essential left ideal of R if $A \cap B \neq (0)$ for every non-zero left ideal B of R.

2·7. LEMMA. Let R be a non-zero semi-prime ring with the ascending chain condition on annihilator left ideals. If Rx and Ry are essential then Rxy is essential.

Proof. Let A be a non-zero left ideal of R and let $\bar{A} = \{r \in R \mid ry \in A\}$. Clearly, $\bar{A}$ is a left ideal of R. Also, $\bar{A}y = Ry \cap A \neq (0)$ by the essentiality of Ry. Thus, $\bar{A} \supseteq \ell(y)$, $\bar{A}y \neq (0)$ and $\ell(y)\cdot y = (0)$; so, $r(\bar{A}) \subsetneqq r\ell(y)$. By lemma 2·6, there exists a non-zero left ideal $T \subseteq \bar{A}$ such that $T \cap \ell(y) = (0)$. If $\bar{T} = \{r \in R \mid rx \in T\}$ then $\bar{T}x = Rx \cap T \neq (0)$ by the essentiality of Rx. Now $\bar{T}xy \neq (0)$; for otherwise, $\bar{T}x \subseteq \ell(y)$ so that $\bar{T}x = T \cap \bar{T}x \subseteq T \cap \ell(y) = (0)$, a contradiction. Since

$\bar{T}xy \subseteq Ty \subseteq \bar{A}y \subseteq A$, we have $Rxy \cap A \neq (0)$. Hence Rxy is essential. This proves the lemma.

2·8. LEMMA. Let R be a non-zero semi-prime ring with the ascending chain condition on annihilator left ideals. If Ra is essential then a is regular in R.

Proof. That $r(a) = (0)$ follows by taking $A = R$ and $B = Ra$ in lemma 2·6. By 2·7, Ra^n is essential for every $n \in Z^+$ and by 2·4, $Ra^n \cap \ell(a) = (0)$ for some n; so, $\ell(a) = (0)$. This proves the lemma.

Until further notice, R is a non-zero semi-prime left Goldie ring.

2·9. LEMMA. R satisfies the descending chain condition on annihilator left ideals.

Proof. Let $L_1 \supset \ldots\ldots \supset L_n \supset \ldots\ldots$ be a properly descending chain of annihilator left ideals of R. By 2·4 and 2·6, there exist non-zero left ideals $C_n \subseteq L_n$ such that $C_n \cap L_{n+1} = (0)$. Since the sum ΣC_n is direct, the chain of annihilator left ideals must terminate.

2·10. LEMMA. If $\ell(c) = (0)$ then Rc is essential and c is regular in R.

Proof. Let A be a non-zero left ideal of R. If $A \cap Rc = (0)$ then $\{Ac^n : n \in Z^+\}$ form a direct sum, a contradiction. Thus $A \cap Rc \neq (0)$. This shows that Rc is essential; 2·8 now shows that c is regular in R.

It may be worthwhile to point out that right regular elements in R need not be regular.

2·11. LEMMA. (a) Every non-zero minimal annihilator ideal of R is a prime left Goldie ring.

(b) There is a finite direct sum of non-zero minimal annihilator ideals of R which is an essential left ideal of R .

Proof. (a) Let S be a non-zero minimal annihilator ideal of R . If T is a non-zero left ideal of S then ST is a left ideal of R contained in T . $ST \neq (0)$, for otherwise $(0) = S(T + TR) \supseteq (T + TR)^2$ which gives $T = 0$ since R is semi-prime. It is now clear that S has no infinite direct sums of non-zero left ideals. Since the ascending chain condition on annihilator left ideals is inherited by subrings, it follows that S is a non-zero left Goldie ring.

Let A,B be ideals of S such that $A \neq (0)$ and $AB = (0)$. Since $SB \subseteq B$, we have $ASB = (0)$; so, $A \subseteq \ell_R(SB) \cap S$. Since SB is a left ideal of R , by 2·4, $\ell_R(SB) \cap S$ is a non-zero annihilator ideal of R contained in S . Minimality of S shows that $S \subseteq \ell_R(SB)$ so that $(SB)^2 = (0)$. Since R is semi-prime, we get $SB = (0)$. Now, $(RB)^2 = (0)$ so $RB = (0)$. This makes B a nilpotent left ideal of R ; so, $B = (0)$. Hence S is a prime ring.

(b) Existence of a non-zero minimal annihilator ideal follows from 2·9. Let $A = S_1 \oplus \cdots \oplus S_n$ be a maximal direct sum of non-zero minimal annihilator ideals of R . Suppose K is a non-zero left ideal of R with $A \cap K = (0)$. Since $AK \subseteq A \cap K$, we have $K \subseteq \mathfrak{r}_R(A)$. Since R is semi-prime, $A \cap \mathfrak{r}_R(A) = (0)$ so that $\mathfrak{r}_R(A)\, A = (0)$. This gives $(0) \neq \mathfrak{r}_R(A) \subseteq \ell_R(A)$. Since R is semi-prime, we have $A \cap \ell_R(A) = (0)$. We can thus find a non-zero minimal annihilator ideal whose intersection with A is zero, thus contradicting our choice of A . This shows that A is an essential left ideal of R . The lemma is now proved.

2·12. LEMMA. If I is an essential left ideal of R then it contains a regular element of R .

Proof. Firstly, we shall prove the lemma in the case when R is a non-zero prime left Goldie ring. Choose $a \in I$ such that $\ell_R(a)$ is minimal in the set $\{\ell_R(x) \mid x \in I\}$. If possible, let J be a non-zero left ideal of R such that $Ra \cap J = (0)$. Since I is an essential left ideal of R, $I \cap J \neq (0)$. We may thus assume that $Ra \cap J = (0)$ and $(0) \neq J \subseteq I$. Let $x \in J$. If $r \in \ell(a+x)$ then $ra = -rx \in Ra \cap J = (0)$. Thus, $\ell(a+x) = \ell(a) \cap \ell(x)$. Minimality of $\ell(a)$ implies $\ell(x) \supseteq \ell(a) \cap \ell(x) \supseteq \ell(a)$; so, $\ell(a)x = (0)$. We have thus shown that $\ell(a)J = (0)$. Since R is assumed to be a prime ring and $J \neq (0)$, this forces $\ell(a) = (0)$. By 2·10, we have $J = (0)$, a contradiction. Hence, Ra is essential and, by 2·10, a is regular in R.

We now prove the lemma for an arbitrary non-zero semi-prime left Goldie ring R. Let $A = S_1 \oplus \cdots \oplus S_n$ be a direct sum of non-zero minimal annihilator ideals of R such that A is an essential left ideal of R; the existence of A was proved in 2·11(b). We claim that $I \cap S_i$ is an essential left ideal of the ring S_i for each i, $1 \leq i \leq n$. If not, let K be a non-zero left ideal of S_i such that $I \cap S_i \cap K = I \cap K = (0)$. Since $S_iK \subseteq K$, we have $I \cap S_iK = (0)$. Essentiality of I implies $S_iK = 0$; so, $K^2 = (0)$. By 2·11(a), S_i is a prime ring. Thus $K^2 = (0)$ implies $K = (0)$, a contradiction. The first part of the proof now shows that $I \cap S_i$ contains an element r_i which is regular in the ring S_i. We claim that $r = r_1 + \dots + r_n$ is a regular element in R. Suppose $\ell_R(r) \neq (0)$. Since A is an essential left ideal of R, $\ell_R(r) \cap A \neq (0)$. Let $0 \neq t = t_1 + \cdots + t_n \in \ell_R(r) \cap A$, where $t_i \in S_i$ for $1 \leq i \leq n$. Since $tr = \sum_{i=1}^{n} t_i r_i = 0$, each $t_i r_i = 0$; so, each $t_i = 0$ by the regularity of r_i in S_i, This gives $t = 0$, contrary to our choice. Thus, $\ell_R(r) = (0)$. 2·10 shows that r is regular in R. This completes the proof.

Observe that, by 2·11 and 2·12, R contains at least one regular element.

We now abandon the restrictions on R and finish the proofs of Goldie's theorems.

2·13. Proofs of Goldie's theorems. 'Only if' parts of both theorems are proved in 2·3.

Firstly, we shall prove the 'if' part of Goldie's second theorem. Let R be a non-zero semi-prime left Goldie ring. As indicated after 2·12, R contains at least one regular element. Let $a, b \in R$ with a regular in R. It is straightforward to check that $I = \{r \in R \mid rb \in Ra\}$ is an essential left ideal of R. By 2·12, I contains a regular element, say c ; so $cb = da$ for some $d \in R$. Thus R has the left common multiple property. By Ore's theorem, R has a total l.q. ring which we denote by Q.

Let L be a non-zero left ideal of Q. Then $L \cap R$ is a left ideal of R. Using the common denominator theorem, we have $L = Q(L \cap R)$. Let K be a left ideal of R such that $(L \cap R) \oplus K$ is an essential left ideal of R ; such a left ideal K exists since R does not have infinite direct sums of non-zero left ideals. Since Q is the total l.q. ring of R, using 2·12, it follows that $Q = L \oplus QK$. Hence Q is a semi-simple Artinian ring. This proves Goldie's second theorem.

Now suppose that R is a non-zero prime left Goldie ring. Since a prime ring is evidently semi-prime, Goldie's second theorem shows that R has a semi-simple Artinian total l.q. ring Q. If Q is not simple, there exist non-zero ideals A,B of Q such that $AB = (0)$. It follows that $A \cap R$ and $B \cap R$ are non-zero ideals of R with $(A \cap R)(B \cap R) = (0)$; however, this is impossible since R is a prime ring. Hence Q must be a simple Artinian ring. This concludes the proof of Goldie's first theorem.

We finish this section with the following.

2·14. PROPOSITION. (a) If R is a left order in a simple (resp. semi-simple) Artinian ring Q then $M_n(R)$ is a left order in the simple (resp. semi-simple) Artinian ring $M_n(Q)$ for every $n \in Z^+$.

(b) If R is a left order in a semi-simple Artinian ring Q and if S is a subring of Q containing R then S is a left order in Q .

(c) If R_i is a left order in a semi-simple Artinian ring Q_i for $1 \leq i \leq n$ then $\bigoplus_{i=1}^{n} R_i$ is a left order in the semi-simple Artinian ring $\bigoplus_{i=1}^{n} Q_i$.

Proof. Observe that if R is a subring of a semi-simple Artinian ring Q such that every $q \in Q$ can be expressed as $q = a^{-1}r$ with $a, r \in R$ then R is a left order in Q ; for, if c is regular in R and $a^{-1}rc = 0$ then $a^{-1}r = 0$; so, $\ell_Q(c) = (0)$ and c is a unit in Q . The proof follows from this observation.

References.

Goldie's theorems in the form in which we have stated them appear in Goldie [2]. Goldie [1] and Lesieur and Croisot [1] contain some special cases of these theorems. By now, quite a few different proofs of Goldie's theorems are available.

We have closely followed Procesi and Small [1] in proving the non-trivial part of Goldie's theorems. Lemma 2·3 is in Goldie [2]. For other proofs of Goldie's theorems, see Faith [1], Goldie [1,2, 4,5], Jacobson [3, Appendix B] and references given there.

Some interesting information concerning semi-prime left Goldie rings is contained in the Faith-Utumi theorem. We have not included the theorem in this monograph only because we have no occasion to use it (and also because a very instructive proof is available in Lambek [1]). See also Faith and Utumi [1] and Jacobson [3, appendix B]. A generalization of the Faith-Utumi theorem for left orders in left Artinian rings is given in Jategaonkar [8].

§3. NIL SUBRINGS.

There are some classical theorems which show that nil subrings of suitably conditioned rings are nilpotent. In this section, we shall prove a very general theorem of this sort due to R.C. Shock.

3·1. LEMMA. Let R be an arbitrary ring and let xy be nilpotent for some $x,y \in R$. If $xyx \neq 0$ then $\ell(x) \subset \ell(xyx)$.

Proof. Let $xyx \neq 0$, $(xy)^k = 0$ and $(xy)^{k-1} \neq 0$. If $(xy)^{k-1} x \neq 0$ then $(xy)^{k-1}(xyx) = 0$ shows that $\ell(x) \subset \ell(xyx)$. If $(xy)^{k-1} x = 0$ then $(xy)^{k-2}(xyx) = 0$ but $(xy)^{k-2} x \neq 0$, so $\ell(x) \subset \ell(xyx)$.

In an arbitrary ring R, the smallest left ideal of R containing a is denoted as $R^1 a$; thus $R^1 a = Ra + Za$.

3·2. SHOCK'S THEOREM. Let R be a ring satisfying the ascending chain condition on left annihilators. Suppose N is a nil subring of R which is not nilpotent. Then there exists a sequence $\{a_n : n \in Z^+\}$ of non-zero elements of N such that

1) $\ell(\{a_n : n \geq k\}) \subsetneqq \ell(\{a_n : n \geq k+1\})$ for every $k \in Z^+$.

2) Either $Ra_n \neq 0$ for every $n \in Z^+$ and $\Sigma_{n\in Z^+} Ra_n$ is direct or $Ra_n = 0$ for every $n \in Z^+$ and $\Sigma_{n \in Z^+} R^1 a_n$ is direct.

Proof. Since R has the a.c.c. on left annihilators, there exists $t \in Z^+$ such that

$$\ell_R(N^t) = \ell_R(N^{t+j}), \quad j \geq 0 .$$

Put $\ell_R(N^t) = K$. Since N is non-nilpotent, $N \nsubseteq K$; so, there exists $x \in N$ such that $R^1 x \nsubseteq K$. Choose $x_1 \in N$ such that $\ell_R(x_1)$ is maximal in the set

$$\{\ell_R(x) \mid x \in N,\ R^1 x \nsubseteq K\} .$$

If $R^1xN \subseteq K$, then $R^1xN^{t+1} = (0)$, so $R^1x \subseteq \ell_R(N^{t+1}) = K$, contrary to our choice of x_1 . Thus, there exists $x \in N$ such that $R^1x_1x \nsubseteq K$. Choose $x_2 \in N$ such that $\ell_R(x_2)$ is maximal in the set

$$\{\ell_R(x) \mid x \in N \ , \ R^1x_1x \nsubseteq K\} \ .$$

Inductively, we can now construct a sequence $\{x_n \colon n \in Z^+\}$ of elements of N such that $\ell_R(x_k)$ is maximal in the set

$$\{\ell_R(x) \mid x \in N \ , \ R^1x_1 \cdots x_{k-1}x \nsubseteq K\}$$

for every $k \in Z^+$.

We need two observations before we come to the crucial part of the proof.

First observation:

$$\ell_R(x_{j+1} \cdots x_{j+m}) \cap Rx_j = (0) \ , \ \forall \ j,m \in Z^+ \ .$$

For, if $rx_j \neq 0$ and $rx_jx_{j+1} \cdots x_{j+m} = 0$ then $\ell_R(x_j) \subset \ell_R(x_jx_{j+1} \cdots x_{j+m})$; the inductive choice of the sequence $\{x_n\}$ shows that

$$R^1x_1 \cdots x_jx_{j+1} \cdots x_{j+m} \nsubseteq K \ .$$

We thus have arrived at a contradiction with the maximality of $\ell_R(x_j)$ in the set

$$\{\ell_R(x) \mid x \in N \ , \ R^1x_1 \cdots x_{j-1}x \nsubseteq K \} \ .$$

We now put

$$a_n = x_1x_2 \cdots x_n \ , \quad n \in Z^+ \ .$$

Second observation:

$$\ell_R(a_1) = \ell_R(a_n) \ , \quad n \in Z^+ \ .$$

Since $a_n = a_1x_2 \cdots x_n$, evidently, $\ell_R(a_1) \subseteq \ell_R(a_n)$. The inductive choice of the sequence $\{x_n\}$ shows that

$$R^1 a_n \not\subseteq K .$$

Thus, if $\ell_R(a_1) \subset \ell_R(a_n)$, we have a contradiction with the maximality of $\ell_R(a_1) = \ell_R(x_1)$ in the set

$$\{\ell_R(x) \mid x \in N , R^1 x \not\subseteq K\} .$$

We now claim that $a_{n+j} x_n = 0$ for all $n \geq 1$ and $j \geq 0$. Assume for a moment that $a_{n+j} x_n \neq 0$ for some $n \geq 1$ and $j \geq 0$. Since $R^1 a_{n+j} x_n \subseteq R x_n$, the first observation above shows that

$$R^1 a_{n+j} x_n x_{n+1} \cdots x_{n+m} \neq 0 , \ \forall \ m \in Z^+ .$$

so, $R^1 a_{n+j} x_n \not\subseteq K$. Since

$$0 \neq a_{n+j} x_n = x_1 \cdots x_{n-1} (x_n x_{n+1} \cdots x_{n+j} x_n)$$

lemma 3·1 shows that if $j > 0$ then

$$\ell_R(x_n) \subset \ell_R(x_n x_{n+1} \cdots x_{n+j} x_n) .$$

Since $R^1 a_{n-1}(x_n x_{n+1} \cdots x_{n+j} x_n) = R^1 a_{n+j} x_n \not\subseteq K$, we obtain a contradiction with our choice of x_n . So, $j = 0$. Now $R^1 a_n x_n = R^1 a_{n-1} x_n^2 \not\subseteq K$ and $\ell_R(x_n) \subseteq \ell_R(x_n^2)$ implies $\ell_R(x_n) = \ell_R(x_n^2)$. However, since x_n is nilpotent, this is possible only if $x_n = 0$, contrary to our choice. This establishes our claim.

We are now in a position to finish the proof. Since $a_{n+j} x_n = 0$ for $j \geq 0$ and $a_{n-1} x_n \neq 0$, it follows that, for every $k \geq 1$,

$$\ell(\{a_n : n \geq k\}) \subsetneq \ell(\{a_n : n \geq k+1\}) .$$

Suppose every $R a_n \neq (0)$. Let

$$r_{s+1} a_{s+1} = \sum_{i=1}^{s} r_i a_i ,$$

where $r_i \in R$ for $1 \leq i \leq s + 1$. Multiplying both sides on the right by x_2 and using the claim proved above, we have $r_1 a_1 x_2 = 0$ i.e. $r_1 \in \ell_R(a_2)$. By the second observation $r_1 \in \ell_R(a_1)$ so $r_1 a_1 = 0$. Now multiplying both sides on the right by x_3, we obtain $r_2 a_2 = 0$. Continuing in this fashion, we conclude that $r_{s+1} a_{s+1} = 0$. Thus $\sum_{n \in Z^+} Ra_n$ is direct.

If some $Ra_k = (0)$ then $Ra_n = 0$ for every $n \geq k$. By choosing k large enough, we may assume that the orders of $a_n, n \geq k$, as elements of the additive abelian group $(R,+)$ are equal (finite or infinite). Now, suppose that

$$n_{s+1} a_{s+1} = \sum_{i=k}^{s} n_i a_i$$

where $n_i \in Z$ for $k \leq i \leq s+1$. Multiplying both sides on the right by x_{k+1}, we have $n_k a_{k+1} = 0$ so $n_k a_k = 0$. Continuing in this fashion, we conclude that $n_{s+1} a_{s+1} = 0$. Thus $\sum_{n \geq k} R^1 a_n$ is direct. Reindexing, if necessary, we have $Ra_n = 0$ for all $n \in Z^+$ and $\sum_{n \in Z^+} R^1 a_n$ is direct. (Observe that the first assertion of the theorem remains valid even after reindexing). This completes the proof.

3·3. COROLLARY (HERSTEIN-SMALL). If R satisfies the ascending chain condition on annihilator left ideals and annihilator right ideals then every nil subring of R is nilpotent.

3·4. COROLLARY (LANSKI). Every nil subring of a left Goldie ring is nilpotent.

References.

Nilpotency of suitably conditioned nil rings is an old affair. A classical theorem of Hopkins' shows that every one-sided nil ideal of a left Artinian ring is nilpotent. Another classical result shows that nil subrings of simple Artinian rings are nilpotent; combining this with Hopkins' theorem, it follows that nil subrings of left Artinian rings are nilpotent.

Generalizing Hopkins' theorem, Levitski showed that the prime radical of a left Noetherian ring is nilpotent. Goldie [2] observed that combining Levitski's theorem with Goldie's second theorem, it follows that nil subrings of left Noetherian rings are nilpotent. An easy proof of Levitski's theorem is given by Utumi [1].

Recent contributions to this area are Herstein and Small [1,2], Levitski [1], Lanski [1], Shock [1,2].

§4. SMALL'S THEOREM.

In this section, we shall prove a theorem due to L.W. Small which characterizes left orders in left Artinian rings.

A few definitions first. Let R be a left Goldie ring. R is said to satisfy the regularity condition if an element $c \in R$ is regular in R whenever $c+P(R)$ is regular in $R/P(R)$. R is said to satisfy the full regularity condition if an element $c \in R$ is regular in R if and only if $c+P(R)$ is regular in $R/P(R)$.

A ring R is called a left T-Goldie ring if R is a left Goldie ring and R/T_k is a left Goldie ring for every $k \in Z^+$, where $T_k = P(R) \cap \mathfrak{r}_R([P(R)]^k)$.

We now state Small's theorem.

4·1. SMALL'S THEOREM. The following conditions on a ring R are equivalent:

(1) R is a left order in a left Artinian ring.

(2) R is a non-nilpotent left T-Goldie ring satisfying the regularity condition.

(3) R is a non-nilpotent left T-Goldie ring satisfying the full regularity condition.

We shall give a cyclic proof: (1) => (3) => (2) => (1). Trivially, (3) => (2). We now prove some lemmas which are needed in both the remaining implications.

Until further notice, R denotes a ring subject to the following restrictions: (i) R is a non-nilpotent left Goldie ring satisfying the regularity condition; (ii) $R/P(R)$ is a left Goldie ring. Let

$$M = \{c \in R \mid c+P(R) \text{ is regular in } R/P(R)\}.$$

Since $R/P(R)$ is a non-zero semi-prime left Goldie ring, Goldie's second theorem shows that $R/P(R)$ contains at least one regular element. The regularity condition implies that M is a semi-group of regular elements in R. We put $T_k = P(R) \cap r_R(\{P(R)\}^k)$ and $R_k = R/T_k$ for every $k \in Z^+$. If $x \in R$, we write $x + T_k$ as x_k.

<u>4.2. LEMMA</u>. $T_{k+n}/T_k = P(R_k) \cap r_{R_k}(\{P(R_k)\}^n)$ for every $k, n \in Z^+$

<u>Proof</u>. Since $T_k \subseteq P(R)$, we have $P(R_k) = P(R)/T_k$. If $x_k \in P(R_k) \cap r_{R_k}(\{P(R_k)\}^n)$ then $x \in P(R)$ and $\{P(R_k)\}^n x_k = 0$ i.e., $\{P(R)\}^n x \subseteq T_k \subseteq r_R(\{P(R)\}^k)$. Thus, $x \in P(R)$ and $\{P(R)\}^{k+n} x = (0)$; so $x \in T_{k+n}$ and $x_k \in T_{k+n}/T_k$. If $y_k \in T_{k+n}/T_k$ then $y \in T_{k+n}$ since $T_{k+n} \supseteq T_k$. So, $\{P(R)\}^{k+n} y = 0$ which gives $\{P(R)\}^n y \subseteq P(R) \cap r_R(\{P(R)\}^k) = T_k$. Thus, $y_k \in P(R_k) \cap r_{R_k}(\{P(R_k)\}^n)$. This proves the lemma.

<u>4.3. LEMMA</u>. If $a \in M$ then $\ell_{R_k}(a_k) = 0$ for every $k \in Z^+$.

<u>Proof</u>. Let $x_k a_k = 0$. Then $xa \in T_k$ so that $xa \in P(R)$ and $\{P(R)\}^k xa = (0)$. Since $a + P(R)$ is regular in $R/P(R)$, $xa \in P(R)$ implies $x \in P(R)$. Since a is regular in R, $\{P(R)\}^k xa = (0)$ yields $\{P(R)\}^k x = (0)$. So, $x \in P(R) \cap r_R(\{P(R)\}^k) = T_k$ and $x_k = 0$. This proves the lemma.

<u>4.4. LEMMA</u>. Assume that M is a left divisor set in R. Let Q be the l.q. ring of R w.r.t. M. Then

(i) $P(Q)$ is nilpotent; $P(Q) \cap R = P(R)$.

(ii) $\{P(Q)\}^n = Q\{P(Q)\}^n = Q\{P(Q) \cap R\}^n$ for every $n \in Z^+$.

(iii) $QT_k \cap R = T_k$ for every $k \in Z^+$.

(iv) QT_k is a two-sided ideal of Q and Q/QT_k is isomorphic with the l.q. ring of R_k w.r.t. M_k where $M_k = \{c + T_k : c \in M\}$.

<u>Proof</u>. We shall firstly show that $QP(R)$ is a nilpotent left ideal of Q. Using the common denominator theorem, it is easily seen that every element of $QP(R)$ is of the form $a^{-1}r$, where $a \in M$ and $r \in P(R)$. If $b \in M$ and $y \in P(R)$ then there exist $c \in M$ and $z \in R$ such that $yb^{-1} = c^{-1}z$ i.e. $cy = zb$; regularity of $b + P(R)$ in $R/P(R)$

shows that $z \epsilon P(R)$. By Shock's theorem, $P(R)$ is nilpotent, say $\{P(R)\}^s = (0)$. To show that $\{QP(R)\}^s = (0)$, it suffices to show that

$$a_1^{-1}r_1a_2^{-1}r_2 \cdots a_s^{-1}r_s = 0 ,$$

where $a_i \epsilon M$ and $r_i \epsilon P(R)$ for $1 \leq i \leq s$. However, this follows easily from what has been said above. (Start from the right; $r_{s-1}a_s^{-1} = c_s^{-1}z_{s-1}$, where $c_s \epsilon M$ and $z_s \epsilon P(R)$; continue). Hence $QP(R)$ is a nilpotent left ideal of Q .

Evidently, $P(Q) \cap R$ is a nil ideal of R so that by Shock's theorem, it is a nilpotent ideal of R . It follows that $P(Q) \cap R \subseteq P(R)$. Since $QP(R)$ is nilpotent, we have $Q\{P(Q) \cap R\} \subseteq QP(R) \subseteq P(Q)$. However, by using the common denominator theorem, it is readily seen that $P(Q) = Q\{P(Q) \cap R\}$. Thus, $P(Q) = QP(R) = QP(R)Q = Q\{P(Q) \cap R\} = Q\{P(Q) \cap R\} Q$. Thus $P(Q)$ is nilpotent. Also, $\{P(Q)\}^2 = QP(R)QP(R) = Q\{P(R)\}^2$, etc.. This proves (i) and (ii).

We now prove (iii). Let $x \epsilon QT_k \cap R$; so $x = c^{-1}r$, where $c \epsilon M$ and $r \epsilon T_k$. Since $T_k \subseteq P(R)$, we have $QT_k \cap R \subseteq QP(R) \cap R \subseteq P(Q) \cap R \subseteq P(R)$. Thus $x \epsilon P(R)$. Since $cx = r \epsilon T_k \subseteq \ell_R(\{P(R)\}^k)$, we have $\{P(Q)\}^k x = Q\{P(R)\}^k Qc^{-1}r = Q\{P(R)\}^k Qr = Q\{P(R)\}^k r = (0)$ by a repeated use of (ii). Thus $(0) = \{P(Q)\}^k x = Q\{P(R)\}^k x = \{P(R)\}^k x$ so that $x \epsilon \ell_R(\{P(R)\}^k)$. Hence $x \epsilon T_k$ and $QT_k \cap R \subseteq T_k$; the other inclusion is clear.

We now prove (iv). Let $a^{-1}r_1 \epsilon QT_k$ and $b^{-1}r_2 \epsilon Q$ where $a, b \epsilon M$, $r_1 \epsilon T_k$ and $r_2 \epsilon R$. We claim that $a^{-1}r_1b^{-1}r_2 \epsilon QT_k$. Since M is assumed to be a left divisor set in R , $r_1b^{-1} = u^{-1}v$ i.e. , $ur_1 = vb \epsilon T_k$, where $u \epsilon M$ and $v \epsilon R$. Thus, $\{P(R)\}^k vb = (0) = \{P(R)\}^k v$ since b is regular in R ; so $v \epsilon \ell_R(\{P(R)\}^k)$. Also, since $vb \epsilon T_k \subseteq P(R)$, $v \epsilon QP(R)Q \cap R = QP(R) \cap R \subseteq P(R)$ by (ii). Hence $v \epsilon T_k$. This proves our claim. It is now clear that QT_k is a two-sided ideal of Q .

We now show that M_k is a left divisor set in R_k. M_k is clearly a semi-group in R_k . If $a_k \in M_k$, $r_k \in R_k$ and $a_k r_k = 0$ then $ar \in T_k$ and $r \in QT_k \cap R = T_k$ by (iii); so $r_k = 0$. Using 4·3 it follows that a_k is regular in R_k . Thus M_k is a semi-group of regular elements in R_k . Let $s_k \in R_k$ and $c_k \in M_k$. Since M is a left divisor set in R and $c \in M$, there exist $c' \in M$ and $s' \in R$ such that $c's = s'c$ i.e., $c'_k s_k = s'_k c_k$ where $c'_k \in M_k$ and $s'_k \in R_k$. Hence M_k is a left divisor set in R_k . By Ore's theorem, R_k has a l.q. ring w.r.t. M_k . It is now straightforward to check that

$$a^{-1}r + QT_k \quad \rightarrow (a + T_k)^{-1} (r + T_k)$$

with $a \in M$, $r \in R$, is an isomorphism of Q/QT_k with the l.q. ring of R_k w.r.t. M_k . This completes the proof.

We now abondon the restrictions on R . The following lemma contains the proof of (1) => (3) in Small's theorem.

4·5. LEMMA. Let R be a left order in a left Artinian ring Q . Then R is a non-nilpotent left T-Goldie ring satisfying the full regularity condition. Further, $P(R) = P(Q) \cap R$; $P(Q) = QP(R)$; $Q/P(Q)$ is isomorphic with the total l.q. ring of $R/P(R)$.

Proof. Since R contains a regular element, R must be a non-nilpotent ring. If ΣA_n is an infinite direct sum of non-zero left ideals of R then, using the common denominator theroem, it is readily seen that ΣQA_n is also an infinite direct sum of non-zero left ideals of Q , which is impossible. Since Q has the ascending chain condition on annihilator left ideals, so does R . Thus R is a left Goldie ring.

We claim that $P(R) = P(Q) \cap R$. (Notice that lemma 4·4 is not available since we have not as yet proved that R has the regularity condition.) By Shock's theorem, $P(R)$ and $P(Q)$ are nilpotent. Thus,

$P(Q) \cap R \subseteq P(R)$. Let $\overline{Q} = Q/P(Q)$. $\overline{Q}$ is a semi-simple Artinian ring. Let $R' = \{r + P(Q) | r \in R\}$. It is easily checked that if $\overline{x} \in \overline{Q}$ then $\overline{x} = c'^{-1}r'$, where $c', r' \in R'$. It follows that $\overline{Q}$ is a total l.q. ring of R' . By Goldie's second theorem, R' is a semi-prime ring. Since $P(R)$ is a nilpotent ideal of R , it now follows that $P(R) \subseteq P(Q) \cap R$. This proves our claim. Consider the map

$$r + P(R) \to r + P(Q)$$

from $R/P(R)$ to R' ; bijectivity of this map follows from $P(R) = P(Q) \cap R$ and it is clearly a homomorphism. Hence $Q/P(Q)$ is isomorphic with the l.q. ring of $R/P(R)$. By Goldie's second theorem, $R/P(R)$ is a semi-prime left Goldie ring.

We now show that R satisfies the full regularity condition. If a is regular in R then it is a unit in Q ; so, $a + P(Q)$ is a unit in $Q/P(Q)$ and thus a regular element in R' . Due to the isomorphism of $R/P(R)$ with R' indicated above, it follows that $a + P(R)$ is regular in $R/P(R)$. If $b + P(R)$ is regular in $R/P(R)$ then $b + P(Q)$ is regular in R'; so, it is a unit in $Q/P(Q)$. Since $P(Q)$ is nilpotent, b is a unit in Q and thus a regular element in R . It is now clear that $M = \{a \in R | a + P(R)$ is regular in $R/P(R)\}$ is precisely the set of all regular elements of R . Since Q is the total quotient ring of R , it follows that M is a left divisor set in R and Q is the l.q. ring of R w.r.t. M .

Now observe that lemma 4.4 is available and shows that R_k has Q/QT_k as a l.q. ring w.r.t. M_k for every $k \in Z^+$. Since Q/QT_k is left Artinian, it is easily seen that R_k is a left Goldie ring for every $k \in Z^+$. Hence R is a left T-Goldie ring. Lemma 4.4 also shows that $P(Q) = QP(R)$. This completes the proof.

It remains to prove the implication (2) => (1) from Small's theorem. In the following three lemmas, R is a non-nilpotent left

T-Goldie ring satisfying the regularity condition. M , T_k and R_k have the meaning defined earlier in this section.

4·6. LEMMA. If $x \in T_k$ and $a \in M$ then there exist $b \in M$ and $y \in R$ such that $bx = ya$.

Proof. The proof is based on an induction on k .

Firstly, observe that since $\ell_R(a) = 0$, Ra is an essential left ideal of R ; for, if A is a non-zero left ideal of R with $A \cap Ra = (0)$ then $\sum_{n \geq 1} Aa^n$ is easily seen to be an infinite direct sum of non-zero left ideals of R , contrary to our hypothesis.

Let $k = 1$. Put $J = \{r \in R \mid rx \in Ra\}$. J is clearly a left ideal of R . Let I be a non-zero left ideal of R . If $Ix = (0)$ then $I \subseteq J$. If $Ix \neq (0)$ then $Ix \cap Ra \neq (0)$ since Ra is an essential left ideal of R . So, there exist $i \in I$ and $r \in R$ such that $ix = ra \neq 0$. If $i \in P(R)$ then, since $x \in T_1 \subseteq \mathfrak{r}_R(P(R))$, $ix = 0$, a contradiction. Thus $i \in J \cap I$ but $i \notin P(R)$; so, $J \cap I \nsubseteq P(R)$. Also, $x \in \mathfrak{r}_R(P(R))$ implies $P(R) \subseteq J$. Summarizing, J is a left ideal of R which properly contains $P(R)$ and meets every non-nilpotent left ideal of R outside $P(R)$. If follows that $J/P(R)$ is an essential left ideal of $R/P(R)$. Since $R/P(R)$ is a non-zero semi-prime left Goldie ring, by lemma 2·12, $J/P(R)$ contains an element $c + P(R)$ which is regular in $R/P(R)$. Since $P(R) \subset J$ and R satisfies the regularity condition, we have $c \in M \cap J$. Thus the lemma is true for $k = 1$.

Now assume the lemma to be true for $k = n$ and let $x \in T_{n+1}$. By 4·3, $\ell_{R_n}(a_n) = (0)$. Consider the set $K = \{r_n \in R_n \mid r_n x_n \in R_n a_n\}$. Recall that $T_{n+1}/T_n = P(R_n) \cap \mathfrak{r}_{R_n}(P(R_n))$. We may then repeat the argument given above for the case $k = 1$; since $P(R_n) = P(R)/T_n$ and $R_n/P(R_n) = R/P(R)$, we obtain $b_n \in K$ for some $b \in M$. Thus, there exists $y \in R$ with $b_n x_n = y_n a_n \in R_n a_n$, i.e., $bx - ya \in T_k$. By the induction hypothesis, there exist $z \in R$ and $c \in M$ such that

$c(bx - ya) = za$, i.e., $cbx = (cy + z)\,a$; clearly, $cb\epsilon M$. This completes the induction on k and concludes the proof.

4·7. LEMMA. M is a left divisor set in R .

Proof. Let $x\epsilon R$ and $a\epsilon M$.

Suppose $x\epsilon P(R)$. By Shock's theorem, $P(R)$ is nilpotent; let s be the index of nilpotency of $P(R)$. If $s = 1$ then $P(R) = (0)$ and $x = 0$; in this case, $a\cdot 0 = 0\cdot a$ is the required left common multiple. If $s > 1$ then $\{P(R)\}^{s-1} \neq (0)$ so $x\epsilon P(R) \cap \ell_R(\{P(R)\}^{s-1}) = T_{s-1}$; lemma 4·6 now gives the required left common multiple.

Suppose $x \notin P(R)$. Since $R/P(R)$ is a non-zero semi-prime left Goldie ring, using Goldie's second theorem or the first part of 2·13, we obtain $b\epsilon M$ and $y\epsilon R$ such that $bx - ya\epsilon P(R)$. As shown above, there exist $c\epsilon M$ and $z\epsilon R$ such that $c(bx - ya) = za$, i.e., $cbx = (cy + z)\,a$. Clearly, $cb\epsilon M$. Due to the regularity condition, M is a semi-group of regular elements of R . This completes the proof.

Ore's theorem and lemma 4·7 show that R has a l.q. ring w.r.t. M . We denote the l.q. ring of R w.r.t. M by Q . Observe that lemma 4·4 is available.

4·8. LEMMA. Q is a left Artinian total l.q. ring of R .

Proof. To show that Q is a left Artinian ring, we shall construct a composition series of left ideals of Q . Let s be the index of nilpotency of $P(R)$. Then $T_{s-1} = P(R)$. By 4·4, $Q/P(Q) \cong Q_{s-1}$, the l.q. ring of R_{s-1} w.r.t. M_{s-1} . However, M_{s-1} is the semi-group of all regular elements of the non-zero semi-prime left Goldie ring R_{s-1} . By Goldie's second theorem, Q_{s-1} is a semi-simple Artinian ring. Thus $Q/P(Q)$ is semi-simple Artinian. Since $Q/QT_k \cong Q_k$, it is easily seen that Q/QT_k has no infinite direct sums of non-zero left ideals. In particular,

QT_{k+1}/Q_k as a left Q-module has no infinite direct sums of Q-submodules. Now, $P(R)T_{k+1} \subseteq T_k$ so that, by 4.4, $QP(R)T_{k+1} \subseteq P(Q)T_{k+1} \subseteq QT_k$. Thus, QT_{k+1}/QT_k may be canonically considered as a left module over $Q/P(Q)$ and as such, it has no infinite direct sums of non-zero submodules. Since $Q/P(Q)$ is a semi-simple Artinian ring, it follows that $Q/P(Q)$ and QT_{k+1}/QT_k have composition series as left Q-modules. These composition series provide a composition series of left ideals of Q. Jordan-Holder theorem now shows that Q is a left Artinian ring.

It remains to show that Q is in fact a total l.q. ring of R. For this, it suffices to show that every regular element in R is a unit in Q. Let c be regular in R. We can express c as $c = a^{-1}r$ for some $a \in M$ and $r \in R$. It readily follows that c is left regular in Q. Consider the descending chain of left ideals $\{Qc^n : n \in Z^+\}$ of Q. Since Q is left Artinian, there exists $n \in Z^+$ such that $Qc^{n+1} = Qc^n$; so $(qc - 1)c^n = 0$ for some $q \in Q$. Since c is left regular in Q, we get $qc = 1$; so $(cq - 1)c = 0$ and $cq = 1$. Hence c is a unit in Q. This completes the proof.

4.9. Proof of Small's theorem. Follows from lemmas 4.5 and 4.8.

References.

Small's theorem was anticipated by Talintyre [1,2]. Small's proof in Small [1] contains an error which was corrected in Gupta and Saha [1] and Small [2]. Some other papers on left orders in left Artinian rings are Gupta [1,2], C.R. Hajarnavis (unpublished), Jans [1], Jategaonkar [8], Mewborn and Winton [1], Robson [1] and Small [3].

CHAPTER II.

STRUCTURE OF PLI-RINGS.

INTRODUCTION.

Our object in this chapter is to develop a structure theory for pli-rings.

In §1, we present some generalities concerning matrix rings. The main results in §2 are Goldie's third theorem and Robson's theorem. These results provide an analogue of the Wedderburn-Artin theorem for semi-prime pli-rings. In §3, we define skew polynomial rings and obtain a generalization of the well-known fact that $R[x]$ is a commutative PIR if and only if R is a field. The significance of our result becomes clear in §§6 and 7.

In §§4-8, we consider non-semi-prime pli-rings. Three theorems due to R. E. Johnson proved in §4 and a theorem due to the author proved in §5 provide a foundation for the study of arbitrary pli-rings. §§6 and 7 contain the structure theory of pli-rings. The results developed so far are used in §8 to prove some structure theorems for pli-rings which satisfy further restrictive conditions.

In the rest of this introduction, we indicate the type of structure theory we have obtained.

Johnson's first theorem shows that every pli-ring is a direct sum of a finite number of primary pli-rings; so, it suffices to consider in detail primary pli-rings. Using Johnson's second theorem and Goldie's third theorem, it is easily seen that every primary pli-ring R is isomorphic with $M_n(S)$ where S is a left Noetherian completely primary ring; some examples indicate that S need not be a pli-ring. However, it is possible to characterize those primary pli-rings R which are isomorphic with $M_n(S)$, where S is a completely primary pli-ring. Although all this generalizes the Wedderburn-Artin theorem, it is not quite satisfactory since it does not provide any clue to the structure of completely primary pli-rings (even left-right Artinian pli-pri-rings).

We are thus lead to search for a new model. Now observe that if K is a field then $K[x]/(x^n)$ is a completely primary PIR. Of course, all completely primary pli-rings can not be so nice. For one thing, $K[x]/(x^n)$ is commutative and so left-right symmetric. Also, it is isomorphic with the graded ring associated with the filtration on it induced by the powers of its prime radical. However, this last observation gives us the clue we want. Given a primary pli-ring R, we consider $G_{P(R)}(R)$, the graded ring associated with the filtration on R induced by the powers of the prime radical $P(R)$ of R and show that $G_{P(R)}(R)$ is almost like $K[x]/(x^n)$; under some mildly restrictive hypothesis, the converse is also valid. It may be worthwhile to point out that our results are new even for left-right Artinian completely primary pli-pri-rings. Also, similar results are obtained for generalized uniserial rings; these will be published elsewhere.

For the rest of this monograph, all rings are assumed to possess unity. All subrings, homomorphisms, modules and bimodules are unitary.

§1. MATRIX RINGS.

The purpose of this section is to gather together some generalities concerning rings of matrices.

Recall that an indexed subset $\{e_{ij}: 1 \leq i, j \leq n\}$ of a ring R is called a system of $n \times n$ <u>matrix units</u> in R if $\sum_{i=1}^{n} e_{ii} = 1$ and $e_{ij}e_{k\ell} = \delta_{jk}e_{i\ell}$, $1 \leq i,j,k,\ell \leq n$, where δ_{jk} is the Kronecker delta. The subring $C = \{r \in R \mid re_{ij} = e_{ij}r, 1 \leq i,j \leq n\}$ is called the <u>centralizer</u> of $\{e_{ij}: 1 \leq i,j \leq n\}$ in R.

<u>1.1. PROPOSITION</u>. Let $\{e_{ij}: 1 \leq i,j \leq n\}$ be a system of $n \times n$ matrix units in a ring R and let C be its centralizer in R. Then $R \cong M_n(C)$.

<u>Proof</u>. See Jacobson [3, page 52].

<u>1.2. PROPOSITION</u>. If $R/P(R) \cong M_n(\bar{A})$ then there exists a ring A such that $R \cong M_n(A)$ and $A/P(A) \cong \bar{A}$.

<u>Proof</u>. Observe that P(A) is a nil ideal of A, so contained in the Jacobson radical of A and that $P(M_n(A)) = M_n(P(A))$. cf. McCoy [1, page 73]. Now the proof of theorem 1, page 55 of Jacobson [3] can be modified to prove the proposition.

If ${}_RM$ is a left R-module, $M^{(n)}$ denotes the direct sum of ${}_RM$ with itself n times.

<u>1.3. PROPOSITION</u>. Let R be a ring. For a left R-submodule N of $R^{(n)}$, let $\tau(N)$ denote the set of all matrices in $M_n(R)$ all whose row-vectors are in N. Then $\tau(N)$ is a left ideal of $M_n(R)$. For a left ideal A of $M_n(R)$, let $\sigma(A)$ denote the set of all those elements of $R^{(n)}$ which occur as row-vectors of A. Then $\sigma(A)$ is a left R-submodule of $R^{(n)}$. Further, $\sigma\tau(N) = N$ and $\tau\sigma(A) = A$.

Proof. Straightforward.

<u>1.4. COROLLARY.</u> $M_n(R)$ is a pli-ring (resp. semi-pli-ring) if and only if every (resp. every finitely generated) left R-submodule of $R^{(n)}$ has a set of generators containing at most n elements.

<u>Proof</u>. Immediate from 1.3.

For the rest of this section, we adopt the following <u>convention</u>: All endomorphisms of left modules will be applied from the right side. The advantage of this convention is that $\mathrm{End}_R(R) \cong R$ rather than R^{op}.

<u>1.5. PROPOSITION</u>. $\mathrm{End}_R(R^{(n)}) \cong M_n(R)$.

<u>Proof</u>. Routine.

Let $\rho: S \to R$ be an isomorphism of rings and let ${}_SA$, ${}_RB$ be left modules. A map $\omega: {}_SA \to {}_RB$ is called a ρ-<u>semi-linear</u> homomorphism if ω is a homomorphism of additive abelian groups $(A,+)$ to $(B,+)$ and

$$(sa)\omega = (s\rho)(a\omega)$$

for all $s \in S$ and $a \in A$. Given a ρ-semi-linear isomorphism $\omega: {}_SA \to {}_RB$, the map $\sigma: \mathrm{End}_S(A) \to \mathrm{End}_R(B)$ defined by

$$\alpha^\sigma = \omega^{-1}\alpha\omega, \quad \alpha \in \mathrm{End}_S(A)$$

is called the isomorphism induced by ω. It is easily seen that σ is indeed an isomorphism of rings.

<u>1.6. THEOREM</u>. For a fixed positive integer n, the following conditions on a ring R are equivalent:

(1) If ${}_RP$ is a left R-module with ${}_RP^{(n)} \cong {}_RR^{(n)}$ then ${}_RP \cong {}_RR$.

(2) If S is a ring and $\sigma: \mathrm{End}_S(S^{(n)}) \to \mathrm{End}_R(R^{(n)})$ is an isomorphism then there exists an isomorphism $\rho: S \to R$ and a ρ-semi-linear isomorphism $\omega: {}_SS^{(n)} \to {}_RR^{(n)}$ such that σ is induced by ω.

<u>Proof</u>. Assume that R satisfies (1). Let S be a ring and

$\sigma:\text{End}_S(S^{(n)}) \to \text{End}_R(R^{(n)})$ be an isomorphism. The image of $\alpha \in \text{End}_S(S^{(n)})$ under σ will be denoted as α^σ. Put $\Lambda = \{1,\ldots,n\}$. Let $\{a_\lambda : \lambda \in \Lambda\}$ be a basis of ${}_S S^{(n)}$. For $\lambda,\mu \in \Lambda$, let $e_{\lambda\mu}$ be the S-linear endomorphism of $S^{(n)}$ defined by

$$a_\lambda e_{\lambda\mu} = a_\mu ; a_\nu e_{\lambda\mu} = 0 \text{ if } \nu \neq \lambda,\ \nu \in \Lambda.$$

Evidently, $\{e_{\lambda\mu} : \lambda,\mu \in \Lambda\}$ is a system of $n \times n$ matrix units in $\text{End}(S^{(n)})$. Put $f_{\lambda\mu} = e^\sigma_{\lambda\mu}$. Then $\{f_{\lambda\mu} : \lambda,\mu \in \Lambda\}$ is a system of $n \times n$ matrix units in $\text{End}(R^{(n)})$. Let $B_\lambda = \text{Im } f_{\lambda\lambda}, \lambda \in \Lambda$. Then $R^{(n)} = \bigoplus_{\lambda\in\Lambda} B_\lambda$. Also,

$$(f_{\lambda\mu}|B_\lambda):B_\lambda \to B_\mu$$

is a R-homormorphism such that $(f_{\lambda\mu}|B_\lambda)(f_{\mu\lambda}|B_\mu) = \text{id.}$ and $(f_{\mu\lambda}|B_\mu)(f_{\lambda\mu}|B_\lambda) = \text{id.}$. It follows that $(f_{\lambda\mu}|B_\lambda):B_\lambda \to B_\mu$ is an R-isomorphism for all $\lambda,\mu \in \Lambda$. Hence, $B_1^{(n)} \cong R^{(n)}$. By (1), we have $B_1 \cong R$.

Choose $b_1 \in B_1$ such that the singleton set $\{b_1\}$ is a basis of B_1. Let $b_\lambda = b_1 f_{1\lambda}, \lambda \in \Lambda$. Then $\{b_\lambda\}$ is a basis of B_λ. Also, $b_\lambda f_{\lambda\mu} = b_\mu$ and $b_\lambda f_{\nu\mu} = 0$ if $\nu \neq \lambda$.

For $s \in S$ and $\lambda \in \Lambda$, let $T_\lambda(s)$ be the unique S-linear endomorphism of $S^{(n)}$ defined by

$$a_\lambda T_\lambda(s) = s a_\lambda ; a_\mu T_\lambda(s) = 0 \text{ if } \mu \neq \lambda.$$

Clearly, $e_{\lambda\lambda}T_\lambda(s)e_{\lambda\lambda} = T_\lambda(s)$; so, $f_{\lambda\lambda}T^\sigma_\lambda(s)f_{\lambda\lambda} = T^\sigma_\lambda(s)$. Thus there exists a unique element in R, say $\rho_\lambda(s) \in R$, such that $b_\lambda T^\sigma_\lambda(s) = \rho_\lambda(s)b_\lambda$. The map $\rho_\lambda : S \to R$ is easily seen to be a monomorphism. To see that ρ_λ is surjective, for $r \in R$ and $\lambda \in \Lambda$, let $U_\lambda(r)$ be the unique R-linear endomorphism of R^n defined by

$$b_\lambda U_\lambda(r) = r b_\lambda ; b_\mu U_\lambda(r) = 0 \text{ if } \mu \neq \lambda.$$

Let $T \in \text{End}(S^{(n)})$ with $T^\sigma = U_\lambda(r)$. Since $f_{\lambda\lambda}U_\lambda(r)f_{\lambda\lambda} = U_\lambda(r)$, we have $e_{\lambda\lambda}T\, e_{\lambda\lambda} = T$. Thus there exists a unique $s \in S$ with $T_\lambda(s) = T$;

so, $\rho_\lambda(s) = r$. To see that $\rho_\lambda = \rho_\mu$ for all $\lambda, \mu \in \Lambda$, observe that $e_{\lambda\mu}T_\mu(s)e_{\mu\lambda} = T_\lambda(s)$; so, for every $s \in S$,

$$\rho_\lambda(s)b_\lambda = b_\lambda T_\lambda^\sigma(s) = b_\lambda f_{\lambda\mu}T_\mu^\sigma(s)f_{\mu\lambda}$$

$$= b_\mu T_\mu^\sigma(s)f_{\mu\lambda} = \rho_\mu(s)b_\lambda.$$

We shall denote this common isomorphism by $\rho:S \to R$.

Let $\omega:{}_S S^{(n)} \to {}_R R^{(n)}$ be the unique ρ-semi-linear isomorphism defined by $a_\lambda\omega = b_\lambda$, $\lambda \in \Lambda$. Let $\varphi:\mathrm{End}_S(S^{(n)}) \to \mathrm{End}_R(R^{(n)})$ be the isomorphism induced by ω and let $\psi = \varphi\sigma^{-1}$. Clearly, ψ is an automorphism of $\mathrm{End}(S^{(n)})$ such that $e_{\lambda\mu}^\psi = e_{\lambda\mu}$ and $T_\lambda^\psi(s) = T_\lambda(s)$. Assume for a moment that $\psi \neq \mathrm{id.}$. Let $\alpha \in \mathrm{End}(S^{(n)})$ with $\alpha^\psi \neq \alpha$. So, there exists $\lambda \in \Lambda$ such that $a_\lambda\alpha \neq a_\lambda\alpha^\psi$. Pick $\mu \in \Lambda$ with $a_\lambda\alpha e_{\mu\lambda} \neq a_\lambda\alpha^\psi e_{\mu\lambda}$. Put $\beta = e_{\lambda\lambda}\alpha e_{\mu\lambda}$. Then $a_\lambda\beta \neq a_\lambda\beta^\psi$ and $a_\nu\beta = 0 = a_\nu\beta^\psi$ for $\nu \neq \lambda$. Since $(\beta|Sa_\lambda):Sa_\lambda \to Sa_\lambda$ is a S-linear homomorphism, there exists $s \in S$ such that $\beta = T_\lambda(s)$; so, $\beta = \beta^\psi$ contrary to $a_\lambda\beta \neq a_\lambda\beta^\psi$. Hence $\psi = \mathrm{id.}$, i.e., σ is induced by ω. We have thus shown that (1)$\Rightarrow$(2).

Assume that R satisfies (2). Let ${}_R P$ be a left R-module such that there exists an R-isomorphism $\eta:P^{(n)} \to R^{(n)}$. For $\gamma \in \mathrm{End}_R(P^{(n)})$, let $\gamma^\xi = \eta^{-1}\gamma\eta$. Then $\xi:\mathrm{End}_R(P^{(n)}) \to \mathrm{End}_R(R^{(n)})$ is an isomorphism of rings. Let $\{g_{\lambda\mu}:\lambda,\mu \in \Lambda\}$ be the usual system of matrix units in $\mathrm{End}_R(P^{(n)})$. Put $h_{\lambda\mu} = g_{\lambda\mu}^\xi$. Then $\{h_{\lambda\mu}:\lambda,\mu \in \Lambda\}$ is a system of matrix in $\mathrm{End}_R(R^{(n)})$. Let

$$S = \{\gamma \in \mathrm{End}_R(R^{(n)}) | h_{\lambda\mu}\gamma = \gamma h_{\lambda\mu}; \lambda,\mu \in \Lambda\}.$$

By 1.1, there exists an isomorphism $\tau':\mathrm{End}_R(R^{(n)}) \to M_n(S)$. Let $\{a_\lambda:\lambda \in \Lambda\}$ be the usual basis of ${}_S S^{(n)}$ and let $\tau'':M_n(S) \to \mathrm{End}_S(S^{(n)})$ be the isomorphism defined by

$$a_\nu((r_{\lambda\mu})\tau'') = \sum_{\lambda=1}^{n} r_{\nu\lambda}\, a_\lambda;\ 1 \leq \nu \leq n.$$

Consider the isomorphism $\sigma:\mathrm{End}(S^{(n)}) \to \mathrm{End}(R^{(n)})$ where $\sigma = (\tau'\circ\tau'')^{-1}$. By our hypothesis, there exists an isomorphism $\rho: S \to R$ and a ρ-semi-linear isomorphism $\omega: {}_S S^{(n)} \to {}_R R^{(n)}$ such that σ is induced by ω. Let $b_\lambda = a_\lambda \omega$. Then $\{b_\lambda : \lambda \in \Lambda\}$ is a basis of ${}_R R^{(n)}$. Also,

$$b_\nu h_{\lambda\mu} = a_\nu(\omega h_{\lambda\mu}\omega^{-1})\omega = (a_\nu h_{\lambda\mu}^{(\tau'\circ\tau'')})\omega$$

$$= \delta_{\nu\lambda}(a_\mu \omega) \qquad = \delta_{\nu\lambda}\, b_\mu,$$

where $\delta_{\nu\lambda}$ is the Kronecker delta. Let $c_\lambda = b_\lambda \eta^{-1}$, $\lambda \in \Lambda$. Then

$$c_\lambda = b_\lambda\, h_{\lambda\lambda}\eta^{-1} = (b_\lambda \eta^{-1})g_{\lambda\lambda} \in P_\lambda,$$

where P_λ is the submodule of $P^{(n)}$ consisting of all those elements of $P^{(n)}$ which have μ^{th} entry zero for all $\mu \neq \lambda$. So, $\{c_\lambda : \lambda \in \Lambda\}$ is a basis of $P^{(n)}$ and $c_\lambda \in P_\lambda$. It follows that $\{c_\lambda\}$ is a basis of P_λ. Hence $P \cong R$. This completes the proof.

A ring R is a <u>PF_n-ring</u> if, for every left R-module ${}_R P$ with $P^{(n)} \cong R^{(n)}$, we have ${}_R P \cong {}_R R$. The following proposition shows that the property PF_n lifts modulo the Jacobson radical.

<u>1.7. PROPOSITION</u>. Let I be a two-sided ideal of R contained in the Jacobson radical of R. If R/I is a PF_n-ring then so also is R.

<u>Proof</u>. Let ${}_R P$ be a left R-module with $P^{(n)} \cong R^{(n)}$. Then $(P/IP)^{(n)} \cong (R/I)^{(n)}$ as left (R/I)-modules. So, $P/IP \cong R/I$ as left R-modules. Since P and R are finitely generated projectives and I is contained in the Jacobson radical of R, it follows, cf. Bass [2, page 90] that $P \cong R$. This completes the proof.

<u>1.8. PROPOSITION</u>. If $M_m(D) \cong M_n(E)$ where D and E are left Ore domains then $m = n$.

<u>Proof</u>. Let K be the total ℓ.q. skew field of D. Then

$M_m(D) \subseteq M_m(K)$. Since $M_n(E)$ contains n orthogonal idempotents, so does $M_m(K)$. Thus, $n \leq m$. Changing roles of D and E, we have $m \leq n$. This completes the proof.

<u>1.9. PROPOSITION</u>. Let D be a left Ore domain and $n \in Z^+$. Then any basis of the left D-module $D^{(n)}$ contains precisely n elements.

<u>Proof</u>. Let K be the total ℓ.q. skew field of D and B be a basis of $D^{(n)}$. If B contains an infinite number of elements then $\mathrm{End}_D(D^{(n)}) \cong M_n(D) \hookrightarrow M_n(K)$ contains an infinite number of orthogonal idempotents, a contradiction. Thus B is a finite set. Let $|B| = m$. Then $D^{(m)} \cong D^{(n)}$, so $M_m(D) \cong M_n(D)$. Proposition 1.8 shows that $m = n$. This completes the proof.

<u>APPENDIX.</u>

We shall show that a part of theorem 1.6 is available for arbitrary free modules over certain rings. The results in this appendix are not used anywhere in this monograph.

Let S be a ring. A left S-module M is a <u>matrically reduced module</u> if, for any positive integer $n > 1$, there does not exist a left S-module L such that $L^{(n)} \cong M$. A ring S is a (<u>left</u>) <u>matrically reduced ring</u> if $_SS$ is a matrically reduced left S-module.

<u>A.1. LEMMA</u>. Let R, S be matrically reduced rings and let $_RA$, $_SB$ be free modules satisfying the following conditions:

(1) If $_SM$ is a matrically reduced left S-module such that $_SB \cong {_SM}^{(I)}$ for some set I then $_SM \cong {_SS}$.

(2) Every matrically reduced direct summand of $_RA$ is finitely generated.

Let $\sigma:\mathrm{End}_R(A) \to \mathrm{End}_S(B)$ be an isomorphism of rings. Then there exists an isomorphism $\rho:R \to S$ and a ρ-semi-linear isomorphism $\omega: {_RA} \to {_SB}$ such that $\alpha^\sigma = \omega^{-1}\alpha\omega$ for every $\alpha \in \mathrm{End}_R(A)$.

Proof. Let $\{a_\lambda : \lambda \in \Lambda\}$ be a basis of ${}_RA$. For each $\lambda, \mu \in \Lambda$, let $e_{\lambda\mu}$ be the R-linear endomorphism of ${}_RA$ defined by $a_\lambda e_{\lambda\mu} = a_\mu$, $a_\nu e_{\lambda\mu} = 0$ if $\nu \neq \lambda$. Clearly, $e_{\lambda\mu} e_{\nu\tau} = \delta_{\mu\nu} e_{\lambda\tau}$, where $\delta_{\mu\nu}$ is the Kronecker delta. Put $f_{\lambda\mu} = e^\sigma_{\lambda\mu}$ so that $f_{\lambda\mu} f_{\nu\tau} = \delta_{\mu\nu} f_{\lambda\tau}$. Let $B_\lambda = (B)f_{\lambda\lambda}$. Since $\{f_{\lambda\lambda} : \lambda \in \Lambda\}$ is an orthogonal set of idempotents, the sum $\sum_{\lambda \in \Lambda} B_\lambda$ is direct. Consider the S-linear maps $(f_{\lambda\mu}|B_\lambda) : B_\lambda \to B$. Since $f_{\lambda\mu} f_{\mu\mu} = f_{\lambda\mu}$, the range of this map is contained in B_μ. Hence $(f_{\lambda\mu}|B_\lambda) : B_\lambda \to B_\mu$ is S-linear. Further,

$$(f_{\lambda\mu}|B_\lambda) \circ (f_{\mu\lambda}|B_\mu) = (f_{\lambda\lambda}|B_\lambda) = \mathrm{id}._{B_\lambda}.$$

It follows that $(f_{\lambda\mu}|B_\lambda) : B_\lambda \to B_\mu$ is an isomorphism.

We now show that $B = \bigoplus_{\lambda \in \Lambda} B_\lambda$. Let $\{c_\gamma : \gamma \in \Gamma\}$ be a basis of B. Let $b \in B$. Then there exist $\gamma_1, \ldots, \gamma_k \in \Gamma$ such that $b \in \sum_{i=1}^{k} Sc_{\gamma_i}$. Let g_i be the S-linear endomorphism of B defined by $c_{\gamma_i} g_i = c_{\gamma_i}$, $c_\gamma g_i = 0$ if $\gamma \neq \gamma_i$. Then $(B)g_i = Sc_{\gamma_i} \cong {}_SS$. Since S is assumed to be a matrically reduced ring, it follows that each $(B)g_i$ is a matrically reduced left S-module, i.e., for any positive integer $n > 1$, there does not exist a set $\{h^{(i)}_{pq} : 1 \leq p,q \leq n\}$ of elements of $\mathrm{End}_S B$ such that

$$h^{(i)}_{pq}\, h^{(i)}_{st} = \delta_{qs}\, h^{(i)}_{pt}$$

and

$$\sum_{p=1}^{n} h^{(i)}_{pp} = g_i .$$

Put $g = \sum_{i=1}^{k} g_i$. Then $bg = b$. Also,

$$g^{\sigma^{-1}} = \sum_{i=1}^{k} g_i^{\sigma^{-1}} ,$$

where $g_i^{\sigma^{-1}}$ are orthogonal idempotents in $\mathrm{End}_R A$ such that, for any positive integer $n > 1$, there does not exist a set $\{H^{(i)}_{pq} : 1 \leq p,q \leq n\}$ of elements of $\mathrm{End}_R(A)$ satisfying

$$H^{(i)}_{pq}\, H^{(i)}_{st} = \delta_{qs}\, H^{(i)}_{pt};$$

and

$$\sum_{i=1}^{n} H_{pp}^{(i)} = g_i^{\sigma^{-1}} .$$

Put $A_i = A\, g_i^{\sigma^{-1}}$. Then A_i is a matrically reduced direct summand of A. By hypothesis, A_i is finitely generated for $i = 1,\ldots,k$. Thus, there exist $\lambda_1,\ldots,\lambda_\ell \in \Lambda$ such that $\sum_{i=1}^{k} A_i \subseteq \sum_{j=1}^{\ell} R\, a_{\lambda_j}$. Let $e = \sum_{j=1}^{\ell} e_{\lambda_j\lambda_j}$. Then $g^{\sigma^{-1}} e = g^{\sigma^{-1}}$ so that $ge^{\sigma} = g$; i.e., $g(\sum_{j=1}^{\ell} f_{\lambda_j\lambda_j}) = g$. Since $bg = b$, it follows that $b \in \Sigma\, B_{\lambda\lambda}$. Hence $B = \bigoplus_{\lambda\in\Lambda} B_\lambda$.

Since R is a matrically reduced ring, each Ra_λ is a matrically reduced R-module. As above, it follows that each B_λ is a matrically reduced S-module. Now our hypothesis shows that ${}_S B_\lambda \cong {}_S S$ for each $\lambda \in \Lambda$.

For a fixed $\lambda_0 \in \Lambda$, let $\{b_{\lambda_0}\}$ be a basis of B_{λ_0}. Let $b_\lambda = b_{\lambda_0} f_{\lambda_0\lambda}, \lambda \in \Lambda$. Then it is clear that $\{b_\lambda : \lambda \in \Lambda\}$ is a basis of B. The rest is now similar to the first part of the proof of Theorem 1.6. This completes the proof of the lemma.

<u>A.2. THEOREM (WOLFSON)</u>. Let R, S be pli-domains, let ${}_R A$, ${}_S B$ be free modules and let $\sigma : End_R(A) \to End_S(B)$ be an isomorphism of rings. Then there exists an isomorphism $\rho : R \to S$ and a ρ-semi-linear isomorphism $\omega : {}_R A \to {}_S B$ such that

$$\alpha^\sigma = \omega^{-1} \alpha\, \omega$$

for every $\alpha \in End_R(A)$.

<u>Proof</u>. It is well-known that over a pli-domain, submodules of free modules are free cf. Cartan and Eilenberg [1, page 13]. Thus the conditions of the lemma A-1 are fulfilled. The theorem is now immediate.

References

Theorem 1.6, proposition 1.7 and lemma A-1 are due to the author and published here for the first time. Theorem A-2 is due to Wolfson [1]. Wolfson's proof is an intricate adoption of some results of R. Baer; our proof is totally different from Wolfson's.

Note that 1.6 is applicable to semi-firs and A-1 is applicable to firs (in the sense of P. M. Cohn).

We have recently constructed some examples to show that given $n > 1$, there exist non-isomorphic left hereditary left Noetherian domains A, B such that $M_n(A) \cong M_n(B)$. See Jategaonkar [9].

§2. SEMI-PRIME LEFT GOLDIE SEMI-PLI-RINGS.

In this section, we develop a structure theory for semi-prime left Goldie semi-pli-rings. The main results of this section are the following two theorems.

2.1. GOLDIE'S THIRD THEOREM. The following conditions on a ring R are equivalent:

(1) R is a semi-prime left Goldie semi-pli-ring.

(2) $R \cong \bigoplus_{i=1}^{n} R_i$, where each R_i is a prime left Goldie semi-pli-ring.

(3) $R \cong \bigoplus_{i=1}^{n} M_{n_i}(D_i)$, where each D_i is a left Ore domain and every finitely generated left D_i-submodule of $D_i^{(n_i)}$ has a set of generators containing at most n_i elements. (cf. proposition 2.11).

Suppose R satisfies the above equivalent conditions. Then the rings R_i in (2) and the rings $M_{n_i}(D_i)$ in (3) are uniquely determined by R up to order and isomorphism. The positive integer n_i is uniquely determined by the ring $M_{n_i}(D_i)$ for $1 \leq i \leq n$.

This theorem has a remarkable resemblance with the Wedderburn-Artin theorem. It may be worthwhile to point out where it differs from the Wedderburn-Artin theorem. If we try to adopt Wedderburn-Artin theorem to semi-prime pli-rings in a naive way, we are led to expect that (a) the domains D_i in (3) of 2.1 should be precisely the pli-domains and that (b) these domains should be uniquely determined up to order and isomorphism by R. However, there exist domains D such that $M_n(D)$ is a prime pli-pri-ring for every $n > 1$ but D is neither a pli-domain nor a pri-domain. The first such example was constructed by R. Swan. As to the uniqueness of the domains D_i, the problem in its full generality is open. We conjecture that these domains are uniquely determined by R.

Apart from unpleasantness, these deviations from the pattern established by the Wedderburn-Artin theorem cause complications and/or obstacles in the structure theory of non-semi-prime pli-rings developed later in this chapter. The following theorem shows that if we impose a mild restriction on the rings under consideration, we get a really good structure theorem.

We need a definition first. A non-zero left module is uniform if any two non-zero submodules have non-zero intersection.

2.2. ROBSON'S THEOREM. The following conditions on a ring R are equivalent:

(1) R is a prime left Goldie semi-pli-ring in which finitely generated uniform left ideals are mutually isomorphic.

(2) $R \cong M_n(D)$, where D is a semi-pli-domain.

If R satisfies the above equivalent conditions then the integer n is uniquely determined by R and the domain D is uniquely determined by R up to an isomorphism.

We now turn to proofs. The following lemma shows that (1) $\Rightarrow$ (2) in 2.1.

2.3. LEMMA. Let R be a semi-prime left Goldie semi-pli-ring. Then R is a direct sum of a finite number of ideals, each of which is a prime left Goldie semi-pli-ring. These ideals are uniquely determined by R up to order.

Proof. By Goldie's second theorem, R is a left order in a semi-simple Artinian ring Q. In Q, there exist orthogonal central idempotents $e_1, \ldots, e_n$ such that each e_i is indecomposible in the centre of Q and $\sum_{i=1}^{n} e_i = 1$.

Let $S = Re_1 \oplus \ldots \oplus Re_n$. Clearly, S is a subring of Q and $R \subseteq S$. Let

$$E = \{r \in R \mid r\, e_i \in R,\ 1 \leq i \leq n\}.$$

Evidently, SE is an ideal of R. Using the common denominator theorem, we get a regular c in R such that $ce_i \in R$ for $1 \leq i \leq n$; thus

$Sc \subseteq R$. Since Sc is easily seen to be a finitely generated left ideal of R, $Sc = Ra$ for some $a \in R$. Since c is a unit in Q, it follows that a is regular in S. Now, $a = bc$ for some $b \in S$. Thus $S = Rb$. Since $b^2 \in S$, there exists $d \in R$ such that $b^2 = db$, i.e., $(b-d)b = 0$. Regularity of b in S yields $b = d \in R$. Hence $S = R$. It is easily seen that each Re_i is a left order in the simple Artinian ring Qe_i; by Goldie's first theorem, Re_i is a prime left Goldie ring. It is now easy to complete the proof.

The following lemma essentially shows that (2) $\Rightarrow$ (3) in Theorem 2.1.

<u>2.4. LEMMA</u>. Let R be a prime left Goldie semi-pli-ring. Then $R \cong M_n(D)$, where D is a left Ore domain such that every finitely generated left D-submodule of $D^{(n)}$ has a set of generators containing at most n elements. The integer n is uniquely determined by R.

<u>Proof</u>. By Goldie's first theorem, R is a left order in a simple Artinian ring Q. Let N be a complete $n \times n$ system of matrix units in Q, i.e., the centralizer of N in Q is a skew field. Since N is a finite set, by the common denominator theorem, there exists a regular element a in R such that $aN \subseteq R$. Let $M = aNa^{-1}$. Then M is a complete system of matrix units in Q and $Ma \subseteq R$.

Clearly, RMa is a finitely generated left ideal of R and contains the regular element a. Thus, $RMa = Rb$ for some regular b in R. Now, $ca = b$ for some $c \in RM$; so, $c = ba^{-1}$ is a unit in Q and $RM = Rc$.

Let $A = \{r \in R \mid Mr \subseteq R\}$. Clearly, $A = \{r \in R \mid Rcr \subseteq R\}$. However, $Rc = RM \supseteq R$; so, $R \supseteq Rc^{-1}$ and $c^{-1} \in R$. Thus, $A = c^{-1}R$. From the definition of A, we have $MA = A$; i.e., $Mc^{-1}R = c^{-1}R$. So, $R = (cMc^{-1})R$ and $cMc^{-1} \subseteq R$. Now, cMc^{-1} is a system of $n \times n$ matrix units in R and a complete system in Q. If D is the centralizer of cMc^{-1} in R, it follows that D is a domain. By 1.1, $R \cong M_n(D)$. If D is not left Ore, then it is not a left Goldie domain; so there exists an infinite set of non-zero left ideals $\{\Lambda_m : m \in Z^+\}$ of D such that $\Sigma \Lambda_m$ is direct; then $\Sigma M_n(\Lambda_m)$ is an infinite direct sum of non-zero left ideals in $M_n(D) \cong R$, a contradiction. 1.4 and 1.8 complete the proof.

2.5. Proof of Theorem 2.1. Follows easily from 2.3 and 2.4.

We now prove some results which contain the proof of Robson's theorem.

2.6. LEMMA. Let $R = M_n(D)$. Every uniform left ideal U of R is R-isomorphic with a uniform left ideal of R in which no matrix has a non-zero entry outside the first column.

Proof. We put an induction on the number, p, of columns with non-zero entries in matrices in U. Observe that exchanging two fixed columns of all matrices in U gives a R-isomorphism of U with a uniform left ideal of R. Thus the lemma is true for $p = 1$. Assume that $p > 1$ and that the lemma is true for $\leq p-1$. Due to the above observation, we may assume that the non-zero columns of U are the first p columns.

If, for each j, $1 \leq j \leq p$, we can find a matrix in U with column $j \neq 0$ and column $k = 0$ for $k \neq j$, then these matrices generate non-zero left subideals of U which form a direct sum, contrary to our hypothesis that U is uniform. Thus there exists $j_0 (1 \leq j_0 \leq p)$ such that there exists no matrix in U with column $j_0 \neq 0$ and column $k = 0$ for $k \neq j$. By exchanging columns, if necessary, we may assume that $j_0 = p$. Thus, for a matrix in U, the first p-1 columns vanish if and only if the matrix is the zero matrix. So, the map $(a_{ij}) \mapsto (a'_{ij})$ given by $a'_{ip} = 0$, $a'_{ij} = a_{ij}$ if $j \neq p$, defines an R-isomorphism of U with a uniform left ideal of R which has $\leq p-1$ columns with non-zero entries in its matrices. The induction hypothesis now gives the required isomorphism. This completes the induction and concludes the proof.

2.7. LEMMA. Let $R = M_n(D)$. The finitely generated uniform left ideals of R are mutually isomorphic if and only if the same is true for D.

Proof. If U is a finitely generated uniform left ideal of R in which no matrix has a non-zero entry outside the first column and if $\bar{U}$ is the set of all entries of all matrices in U then $\bar{U}$ is easily seen to be a finitely generated uniform left ideal of D. If V is a finitely generated uniform left ideal of D and if V^* is the left ideal of R consisting of all matrices (a_{ij}) with $a_{i1} \in V$, $a_{ij} = 0$ if $j \neq 1$, then V^* is a finitely generated uniform left ideal of R. Further, $\bar{U}^* = U$ and $\overline{V^*} = V$.

Now assume that all finitely generated uniform left ideals of D are mutually isomorphic and let I, J be finitely generated uniform left ideals of R. By 2.6 and the above observations, we obtain finitely generated uniform left ideals V_1 and V_2 of D such that $I \cong V_1^*$ and $J = V_2^*$. If $\varphi: V_1 \to V_2$ is a D-isomorphism then $\varphi^*: V_1^* \to V_2^*$ given by $\varphi^*((a_{ij})) = (\varphi(a_{ij}))$ is readily seen to be an R-isomorphism. Thus $I \cong J$.

Assume that all finitely generated uniform left ideals of R are mutually R-isomorphic and let W_1 and W_2 be finitely generated uniform left ideals of D. Then there is an R-isomorphism $\psi: W_1^* \to W_2^*$. It is clear that any matrix of the form $\left(\begin{array}{c|c} a & 0 \\ \hline 0 & 0 \end{array}\right)$, $a \in W_1$, maps into a matrix of the form $\left(\begin{array}{c|c} b & 0 \\ \hline 0 & 0 \end{array}\right)$, $b \in W_2$. The map $\psi: W_1 \to W_2$ defined by $a \mapsto b$ is seen to be a D-isomorphism. This completes the proof.

2.8. LEMMA. Let D be a semi-pli-domain and n be an arbitrary positive integer. Then D is a left Ore domain and a PF_n-ring. Further, every finitely generated left D-submodule of $D^{(n)}$ is free with a basis containing at most n elements.

Proof. Assume for a moment that D is not left Ore. Then there exist non-zero $a, b \in D$ such that $Da \cap Db = (0)$. Let $Da + Db = Dc$. Then there exist $a', b', f, g \in D$ such that $a = a'c$, $b = b'c$ and $c = fa + gb$. So, $a = a'fa + a'gb$; i.e., $(1-a'f)a = a'gb = 0$. Since D is a domain and $b \neq 0$, either $a' = 0$ or $g = 0$. However, if $a' = 0$ then $a = a'c = 0$, a contradiction and if $g = 0$ then $c \in Da$, so $b \in Dc \subseteq Da$, a contradiction again. Hence D must be a left Ore domain.

Let N be a finitely generated left D-submodule of $D^{(n)}$. Let $\psi: D^{(n)} \to D$ be the D-epimorphism which sends an element of $D^{(n)}$ to its first entry. Then $\psi(N)$ is a finitely generated left ideal of D; so $\psi(N) = Dd$ for some $d \in D$. Pick $x \in N$ with $\psi(x) = d$. Then $N = Dx \oplus \ker \psi$ and $\ker \psi$ is a finitely generated left D-module which is isomorphic with a submodule of $D^{(n-1)}$. An induction is now available to conclude that N is free with a basis containing at most n elements.

Let P be a left D-module such that $P^{(n)} \cong D^{(n)}$. Then P is isomorphic with a submodule of $D^{(n)}$. As shown above, P is free with a basis of k elements where $k \leq n$. Then $P^{(n)}$ has a basis of kn elements; so, $D^{(n)}$ has a basis of kn elements. Proposition 1.9 shows that $k = 1$. Thus ${}_DP \cong {}_DD$, i.e., D is a PF_n-ring. This completes the proof.

2.9. Proof of Theorem 2.2. For $(1) \Rightarrow (2)$, use 2.4 and 2.7. For $(2) \Rightarrow (1)$, observe that, if D is a semi-pli-domain then by 2.8, D is a left order in a skew field K; so, by I.2.14, $M_n(D)$ is a left order in $M_n(K)$; Goldie's first theorem implies that $M_n(D) \cong R$ is a prime left Goldie ring. 2.8 and 1.4 show that R is semi-pli-ring. 2.7 shows that finitely generated uniform left ideals of R are mutually isomorphic. The uniqueness assertions follow from 2.8, 1.5, 1.6 and 1.8. This completes the proof.

We finish this section with some propositions.

2.10. PROPOSITION. If D is a commutative domain such that $M_n(D)$ is a pli-ring (resp. semi-pli-ring) for some $n \in Z^+$ then D is a pli-domain (resp. semi-pli-domain).

Proof. Let I be a non-zero finitely generated ideal of D, say $I = \sum_{j=1}^{m} Dd_j$, where each $d_j \neq 0$. Let I^* be the left ideal of $M_n(D)$ generated by b_j^*, where b_j^* is the diagonal matrix with the first diagonal entry d_j and all other diagonal entries 1. Then there exist c^* and x_j^* in $M_n(D)$ such that $I^* = M_n(D)c^*$ and $c^* = \sum_{j=1}^{m} x_j^* b_j^*$. It follows that every element in the first column of c^* is in I so that $\det c^* \in I$. Since $b_j^* = y_j^* c^*$ for some y_j^* in $M_n(D)$,

$$d_j = \det b_j^* = (\det y_j^*)(\det c^*) \in I,$$

for $1 \leq j \leq m$. Hence $\det c^*$ generates I. It is now easy to complete the proof.

2.11. PROPOSITION. Let D be a domain and let $n \in Z^+$. $M_n(D)$ is a pli-ring if and only if, for non-zero left ideals $P_1,\ldots,P_n$ of D, we have

$$P_1 \oplus \ldots \oplus P_n \cong {}_D D^{(n)}.$$

Proof. Let $M_n(D)$ be a pli-ring and let $P_1,\ldots,P_n$ be non-zero left ideals of D. Since $P_1 \oplus\ldots\oplus P_n$ is a left D-submodule of ${}_D D^{(n)}$, by 1.4, it has a set of generators containing at most n elements, say $v_1,\ldots,v_k$, $k \leq n$. We may take

$$v_i = (d_{i1},\ldots,d_{in}), \quad i = 1,\ldots,k,$$

where all $d_{ij} \in D$. Let A be the matrix in $M_n(D)$ whose first k row vectors are $v_1,\ldots,v_k$ and remaining row vectors 0. Let K_i be the left ideal of $M_n(D)$ consisting of matrices with all row-vectors in P_i, $1 \leq i \leq n$. Then

$$K_1 \oplus \ldots \oplus K_n \subseteq M_n(D)A.$$

Since D is a left Noetherian domain, it has a ℓ.q. skew field, say L; then $Q = M_n(L)$ is the ℓ.q. ring of $M_n(D)$. Since $QK_1 \oplus \ldots \oplus QK_n = Q$, it follows then A is regular in $M_n(D)$; so $k = n$. Now, if $d_i' \in D$ with

$$\sum_{i=1}^{n} d_i' v_i = 0,$$

then

$$\begin{pmatrix} d_1' & d_2' & \ldots & d_n' \\ 0 & 0 & \ldots & 0 \\ \cdot & \cdot & \cdot & \cdot \\ 0 & 0 & \ldots & 0 \end{pmatrix} A = (0).$$

Regularity of A in $M_n(D)$ shows that each $d_i' = 0$. Hence $\{v_i : 1 \leq i \leq n\}$ is a basis of $P_1 \oplus \ldots \oplus P_n$; so $P_1 \oplus \ldots \oplus P_n \cong D^{(n)}$.

Conversely, if $P_1 \oplus \ldots \oplus P_n \cong {}_D D^{(n)}$ for all non-zero left ideals $P_1, \ldots, P_n$ of D then evidently each left ideal of D is a finitely generated projective. i.e., D is a left hereditary left Noetherian domain. Let N_1 be a submodule of ${}_D D^{(n)}$. Since ${}_D D^{(n)}$ is Noetherian, we obtain a submodule N_2 such that $N = N_1 \oplus N_2$ is an essential submodule of $D^{(n)}$. By a well-known theorem of Kaplansky, see Cartan-Eilenberg [1, page 13], $N \cong P_1 \oplus \ldots \oplus P_\ell$ where each P_i is a non-zero left ideal of D. Using essentiality of N, it follows that $\ell = n$ (either use existence of simple Artinian ℓ.q. ring of $M_n(D)$ or see Goldie [4, Chapter 3, page 56]). By our hypothesis, N has a system of generators containing at most n elements so N_1 also has a system of generators containing at most n elements. By 1.4, $M_n(R)$ is a pli-ring. This completes the proof.

<u>2.12. COROLLARY</u>. Let D be a domain such that $M_n(D)$ is a pli-ring. Then $M_k(D)$ is a pli-ring for each $k \geq n$.

<u>Proof</u>. Let $P_1, \ldots, P_k$ be non-zero left ideals of D. Then, using 2.11, we have

$$\overset{k}{\underset{i=1}{\oplus}} P_i = (\overset{n}{\underset{i=1}{\oplus}} P_i) \oplus (\overset{k}{\underset{i=n+1}{\oplus}} P_i) \cong D^{(n)} \oplus (\overset{k}{\underset{i=n+1}{\oplus}} P_i)$$

$$= D \oplus (D^{(n-1)} \oplus P_{n+1}) \oplus (\overset{k}{\underset{i=n+2}{\oplus}} P_i)$$

$$= D^{(n+1)} \oplus (\overset{k}{\underset{i=n+2}{\oplus}} P_i) \text{ etc. } \cong D^{(k)}.$$

This completes the proof.

References.

Goldie's third theorem is essentially in Goldie [3]. He shows the equivalence of (1) and (2) and shows that a prime pli-ring is a matrix ring over a left Noetherian domain. The trivial changes needed to make Goldie's third theorem a characterization are believed to be new. Our proof of Goldie's third theorem is taken from Robson [2]; Robson's proof of lemma 2.4 is patterned after the Faith-Utumi theorem.

Robson [2] shows that (1)$\Leftrightarrow$(2) in Theorem 2.2. The observation concerning the uniqueness appears to be new. A result related to Robson's theorem is proved by Cohn [4].

Proposition 2.10 is in Robson [2]; see also Faith-Utumi [1]. So far as we know, 2.11 and 2.12 are new.

§3. SKEW POLYNOMIAL RINGS.

In this section, we shall consider skew polynomial rings of a special type over prime pli-rings. Certain epimorphic images of these rings play a prominent part in describing the structure of arbitrary pli-rings.

Let R be a ring and ρ: $R \to R$ be a monomorphism. Let $R[X,\rho]$ denote the set of all formal polynomials in the indeterminate X with coefficients in R written on the left of power of X. Define equality and addition in $R[X,\rho]$ as usual; define a multiplication by assuming the distributive laws and the rule

$$Xr = \rho(r)X$$

for every $r \in R$. It is straightforward to check that $R[X,\rho]$ is a ring. The ring $R[X,\rho]$ is called a skew polynomial ring over R. A skew power series ring $R[[X,\rho]]$ over R is defined in an analogous way.

The main result of this section is the following.

3.1. THEOREM. Let R be a prime pli-ring and Q be the simple Artinian l.q. ring of R. Let ρ: $Q \to R$ be a monomorphism. Then $R[x,\rho]$ is a prime pli-ring.

Till lemma 3.4, we shall work in the following context. D is a left Noetherian domain with total l.q. skew field K, ρ: $K \to D$ is a monomorphism and $A = D[x,\rho]$; n is a positive integer such that $R = M_n(D)$ is a prime pli-ring. We shall be concerned with the ring $S = M_n(A)$. Observe that, by 1.4, every left ideal of D has a set of generators containing at most n elements.

3.2. LEMMA. If ℓ is a non-zero left ideal of A then there exist a_1, ----, $a_n \in A$, all of the same degree such that $\ell = \sum_{i=1} A a_i$.

Proof. Let $m = \min\{\deg a \mid a \in \mathcal{l}, a \neq 0\}$. If $a = dx^m +$ (lower degree terms) is a non-zero polynomial in $\mathcal{l}$ then $[\rho(d)]^{-1} x\, a = x^{m+1} +$ (lower degree terms) is in $\mathcal{l}$. It follows that $\mathcal{l}$ is generated by the polynomials of degree m belonging to $\mathcal{l}$. If $\bar{\mathcal{l}}$ is the set consisting of zero together with the leading coefficients of all the polynomials of degree m in $\mathcal{l}$ then $\bar{\mathcal{l}}$ is a non-zero left ideal of D. Using 1.4, we get $d_1, ----, d_n \in D^*$ such that $\bar{\mathcal{l}} = \sum_{i=1}^{n} Dd_i$. Pick $a_i \in \mathcal{l}$ with $\deg a_i = m$ and with the leading coefficient d_i. Then $\mathcal{l} = \sum_{i=1}^{n} A\, a_i$. This proves the lemma.

For the sake of convenience, we introduce some ad hoc terminology; it will be used in the next two lemmas only. Let Λ be a non-zero left ideal of S. Let p_1 be the first positive integer such that there exists a matrix in Λ with non-zero p_1^{th} column. The integer p_1 is called the initial column of Λ. Let $\mathcal{l}_{p_1}$ be the set of all entries in the p_1^{th} column of matrices in Λ. Then $\mathcal{l}_{p_1}$ is a non-zero left ideal of A, called the initial left ideal of Λ. Let m_{p_1} be the least degree of non-zero polynomials in $\mathcal{l}_{p_1}$; m_{p_1} is called the initial degree of Λ. An element $(a_{ij}) \in \Lambda$ is called a prominent element in Λ if $\mathcal{l}_{p_1} = \sum_{i=1}^{n} A\, a_{ip_1}$ and $\deg a_{ip_1} = m_{p_1}$ for $1 \leq i \leq n$. Let $\Lambda' = \{\lambda \in \Lambda \mid p_1^{th}$ column of λ is zero$\}$. Clearly, Λ' is a left ideal of S with $\Lambda' \subset \Lambda$; Λ' is called the main left subideal of Λ.

3.3. LEMMA. Every non-zero left ideal Λ of S contains a prominent element.

Proof. Follows immediately from 3.2.

3.4. LEMMA. S is a prime pli-ring.

Proof. It is clear that the degree rule holds in A; so, A is a domain and $S = M_n(A)$ is a prime ring.

Let Λ be a non-zero left ideal of S. There exists a finite chain of left ideals of S

$$\Lambda = \Lambda_1 \supset \Lambda_2 \supset \cdots\cdots \supset \Lambda_k \supset \Lambda_{k+1} = (0)$$

such that $1 \leq k \leq n$ and Λ_{t+1} is the main left subideal of Λ_t for $1 \leq t \leq k$. Let p_t be the initial column of Λ_t, m_{p_t} be the initial degree of Λ_t and ℓ_{p_t} be initial left ideal of Λ_t for $1 \leq t \leq k$. By 3.3, we can choose a prominent element F_t' in Λ_t for each t. Then we can normalize these prominent elements to obtain matrices $F_t = (f_{ij}^{(t)})$ with entries in A such that

(1) F_t is a prominent element in Λ_t for $1 \leq t \leq k$;

(2) $\deg f_{ip_q}^{(t)} \leq m_{p_q}$ if $1 \leq t \leq q \leq k$ and $1 \leq i \leq n$.

Such a normalization is possible since $\rho(d)$ is a unit in D for every non-zero $d \in D$.

For each t, $1 \leq t \leq k$, we define a nxn matrix $(\Pi_{ij}^{(t)}) \in R$ as follows:

(1) $\Pi_{ij}^{(t)} = 0$ if $j \neq p_t, p_{t+1}, \cdots, p_k$;

(2) $\Pi_{ip_q}^{(t)}$ = coefficient of $x^{m_{p_q}}$ in $f_{ip_q}^{(t)}$ for $1 \leq i \leq n$ and $t \leq q \leq k$.

Consider the left ideal $\sum_{t=1}^{k} R(\Pi_{ij}^{(t)})$ of R. Since R is a pli-ring, there exists a nxn matrix (γ_{ij}) with entries in D such that $\sum_{i=1}^{k} R(\Pi_{ij}^{(t)}) = R(\gamma_{ij})$. Thus, there exist matrices $(\alpha_{ij}^{(t)})$ and $(\beta_{ij}^{(t)})$ with entries in D such that

$$(\alpha_{ij}^{(t)})\ (\gamma_{ij}) = (\Pi_{ij}^{(t)}) \quad \text{for } 1 \leq t \leq k;$$

$$\sum_{t=1}^{k} (\beta_{ij}^{(t)})\ (\Pi_{ij}^{(t)}) = (\gamma_{ij}).$$

Consequently, for each t_o, $(1 \leq t_o \leq k)$, we have

$$(\alpha_{ij}^{(t_o)}) \sum_{t=1}^{k} (\beta_{ij}^{(t)}) (\Pi_{ij}^{(t)}) = (\Pi_{ij}^{(t_o)}). \qquad (*)$$

Put $H = \sum_{t=1}^{k} (\beta_{ij}^{(t)}) F_t$ and put

$$H_{t_o} = (\alpha_{ij}^{(t_o)}) H = (\alpha_{ij}^{(t_o)}) \sum_{t=1}^{k} (\beta_{ij}^{(t)}) F_t = (h_{ij}^{(t_o)})$$

for $1 \leq t_o \leq k$. We claim that H_{t_o} is a prominent element in Λ_{t_o} for each t_o. For $1 \leq q \leq t_o$, let $\mathfrak{c}_{p_q}^{(t_o)}$ denote the p_q^{th} column of H_{t_o}. By the definition of H_{t_o}, we have

$$\mathfrak{c}_{p_q}^{(t_o)} = (\alpha_{ij}^{(t_o)}) \sum_{t=1}^{k} (\beta_{ij}^{(t)}) (p_q^{th} \text{ column of } F_t)$$

$$= (\alpha_{ij}^{(t_o)}) \sum_{t=1}^{k} (\beta_{ij}^{(t)}) (\Pi_{ip_q}^{(t)} x^{m_{p_q}} + g_{ip_q}^{(t)})$$

where $f_{ip_q}^{(t)} = \Pi_{ip_q}^{(t)} x^{m_{p_q}} + g_{ip_q}^{(t)}$ with $\deg g_{ip_q}^{(t)} < m_{p_q}$.

Using the equation $(*)$, we obtain $\bar{g}_{ip_q}^{(t_o)} \in A$ with degree $< m_{p_q}$ such that

$$h_{ip_q}^{(t_o)} = \Pi_{ip_q}^{(t_o)} x^{m_{p_q}} + \bar{g}_{ip_q}^{(t_o)},$$

for all $1 \leq i \leq n$ and $1 \leq q \leq k$.

Suppose $t_o = 1$. Then each $h_{ip_1}^{(1)}$ is a polynomial of degree m_{p_1} belonging to $\mathfrak{l}_{p_1}$. Since the leading coefficients of $h_{ip_1}^{(1)}$ and $f_{ip_1}^{(1)}$ are the same and m_{p_1} is the least degree of non-zero polynomials in $\mathfrak{l}_{p_1}$, it follows that $h_{ip_1}^{(1)} = f_{ip_1}^{(1)}$. Hence H_1 is a prominent element in Λ_1.

Suppose $t_o > 1$. Then $\Pi_{ip_1}^{(t_o)} = 0$ for $1 \leq i \leq n$; so, $h_{ip_1}^{(t_o)} = \bar{g}_{ip_1}^{(t_o)}$ and $\deg h_{ip_1}^{(t_o)} < m_{p_1}$. It follows from the definition of m_{p_1} that $h_{ip_1}^{(t_o)} = 0$. Hence $H_{t_o} \in \Lambda_2$. Repeating this argument, we can show that $H_{t_o} \in \Lambda_{t_o}$. Now, as in the preceding paragraph, we get

$h^{(t_0)}_{ip_{t_0}} = f^{(t_0)}_{ip_{t_0}}$. Hence H_{t_0} is a prominent element in Λ_{t_0}. This proves our claim. Since $\{H_t: 1 \leq t \leq k\}$ generate Λ as a left ideal of S, we have $\Lambda = SH$. Hence S is a pli-ring. This completes the proof of the lemma.

3.5. Proof of Theorem 3.1. Clearly $\rho(Q) \subseteq R$ and $\rho(Q)$ is a simple Artinian ring. Choose a complete system of matrix units $\{e_{ij}: 1 \leq i, j \leq n\}$ in $\rho(Q)$. Then $\{\rho(e_{ij}): 1 \leq i, j \leq n\}$ is also a complete system of matrix units in $\rho(Q)$. Since $\rho(Q)$ is a simple Artinian ring, there exists a unit $u \in \rho(Q)$ such that $u\, \rho(e_{ij})u^{-1} = e_{ij}$ for $1 \leq i, j \leq n$. cf. Jacobson [3; page 59]. Let ψ: $Q \to R$ be the monomorphism defined by $\psi(q) = u\, \rho(q)u^{-1}, q \in Q$. If D and K are centralizers of $\{e_{ij}: 1 \leq i, j \leq n\}$ in R and Q respectively then K is the total l.q. sfield of D and ψ induces a monomorphism $\bar{\psi}$: $K \to D$. By 1.1, $R \cong M_n(D)$. Now, $M_n(D[y,\bar{\psi}]) \cong M_n(D)\,[y,\psi]$. If we put $y = ux$, it is readily seen that $M_n(D)\,[y,\psi] \cong M_n(D)\,[x,\rho] \cong R[x,\rho]$. By 3.4, $M_n(D[y,\bar{\psi}])$ is a prime pli-ring. Hence $R[x,\rho]$ is a prime pli-ring. This completes the proof.

We have a few comments on theorem 3.1. Notice that the theorem and its proof can be adopted to skew power series ring $R[[x,\rho]]$. A similar theorem is probably true for prime left Goldie semi-pli-rings. The proof of theorem 3.1 can be adopted if one can show that the main left subideal Λ' of a finitely generated left ideal Λ of S is finitely generated. Some results of Chase [1] appear to be related.

Later on, we shall see that theorem 3.1 has a converse; viz., if $R[x,\rho]$ is a prime pli-ring then R is a prime pli-ring and ρ: $R \to R$ can be extended to a monomorphism ρ: $Q \to R$, where Q is the simple Artinian total l.q. ring of R.

References.

Skew polynomial rings $K[x,\rho,\delta]$ over skew fields K were introduced by Ore [1]. (see §1 of chapter IV for definition). Iterated extensions of this type of domains were considered by Curtis [1]. The rings $K[x,\rho,\delta]$ were extensively studied in case K is a field, ρ is an automorphism and $\delta = 0$ and in case K is a field, δ = id. and $\delta \neq 0$; these results are summarized in Jacobson [2; chapter 3]. Rings $K[x,\rho]$, where K is a field and ρ is a monomorphism, were used mostly as bad examples of rings; see Divinski [1].

The significance of the skew polynomial rings for the structure theory of arbitrary pli-rings was made clear in Jategaonkar [1,5]. In Jategaonkar [3], theorem 3.1 was proved under the hypothesis that R is a pli-domain; a transfinite generalization of it appears in Jategaonkar [4]. Theorem 3.1 in its present form is new. Skew polynomial rings on left orders in left Artinian rings are considered in detail in Jategaonkar [8].

§4. JOHNSON'S THEOREMS.

In this section, we shall prove three theorems of R. E. Johnson. These theorems provide some important information regarding the structure of fully left Goldie ipli-rings and are repeatedly needed in the sequel.

A definition is needed. A ring R is a primary ring if, for ideals A, B of R, $AB = (0)$ implies either $A = (0)$ or $B^n = (0)$ for some positive integer n. A primary ring R is completely primary if $R/P(R)$ is a domain.

4.1. JOHNSON'S FIRST THEOREM. A ring R is a fully left Goldie ipli-ring if and only if R is a direct sum of a finite number of ideals, each of which is a primary fully left Goldie ipli-ring.

4.2. JOHNSON'S SECOND THEOREM. Let R be a fully left Goldie ipli-ring. R is a primary ring if and only if $R/P(R)$ is a prime ring.

4.3. JOHNSON'S THIRD THEOREM. Let R be a primary fully left Goldie ipli-ring. If Rc is a non-nilpotent two-sided ideal of R then $P(R) \subset Rc$ and $\ell(c) = (0)$.

The usefulness of Johnson's first theorem is evident; it reduces the study of arbitrary fully left Goldie ipli-rings to that of primary fully left Goldie ipli-rings. Johnson's second theorem connects the primary fully left Goldie ipli-rings with prime rings of the same kind. However, it is the third theorem which provides the most important information regarding primary fully left Goldie ipli-rings. It shows that the prime radical occupies a peculiar position in such rings; the significance of the left regularity of the generators of non-nilpotent ideals is not so easily explained; loosely stated, it shows that an ideal is either big or very small.

We shall note some useful trivialities before we prove Johnson's theorems. It is clear that a homomorphic image of a fully left Goldie ipli-ring is again a fully left Goldie ipli-ring; also, a finite direct sum of fully left Goldie ipli-rings is a fully left Goldie ipli-ring and conversely.

4.4. PROPOSITION. If R is an arbitrary ring and Ra, Rb are ideals of R then $RaRb = Rab$.

Proof. Evidently, $aR \subseteq Ra$, so that $RaRb \subseteq Rab$. On the other hand, $ab \in RaRb$, so $Rab \subseteq RaRb$.

4.5. PROPOSITION. If R is a ipli-ring then $P(R)$ is nilpotent.

Proof. Let $P(R) = Rw$. Since $P(R)$ is nil, $w^n = 0$ for some positive integer n. By 4.4, $\{P(R)\}^n = Rw^n = (0)$.

Now we shall prove Johnson's theorems.

4.6. LEMMA. Let R be a fully left Goldie ipli-ring such that $R/P(R)$ is a prime ring. Let Rc be a non-nilpotent two-sided ideal of R. Then $P(R) \subset Rc$ and $\ell(c) = (0)$.

Proof. Let $B = Rc + P(R) = Rb$. Then $\bar{B} = \{r + P(R) | r \in B\}$ is a non-zero ideal of $\bar{R} = R/P(R)$. Also, $\bar{B} = \overline{Rb}$, where $\bar{b} = b + P(R)$. Since $\bar{R}$ is a prime ring, using 4.4, we have $\mathfrak{r}_{\bar{R}}(\bar{b}^m) = (\bar{0})$ for every positive integer m. If Λ is a left ideal of $\bar{R}$ with $\Lambda \cap \bar{R}\bar{b}^m = (\bar{0})$, then $\bar{R}\bar{b}^m\Lambda = (0)$ so that $\Lambda \subseteq \mathfrak{r}_{\bar{R}}(\bar{b}^m) = (\bar{0})$. However, by 1.2.5, there exists $m \in Z^+$ such that $\ell_{\bar{R}}(\bar{b}) \cap \bar{R}\bar{b}^m = (\bar{0})$. Thus, $\ell_{\bar{R}}(\bar{b}) = (\bar{0})$; i.e., for $r \in R$, $rb \in P(R)$ if and only if $r \in P(R)$. Since $P(R) \subset Rb$, we obtain $P(R) = P(R)b$; on induction, this gives $P(R) = P(R)b^n \subseteq Rb^n$ for every positive integer n. By lemma 1.2.5, there exists $s \in Z^+$ such that $\ell_R(b) \cap Rb^s = (0)$. Left regularity of $\bar{b}$ in $\bar{R}$ shows that $\ell_R(b) \subseteq P(R)$. It follows that $\ell_R(b) = (0)$.

By 4.5, $\{P(R)\}^t = (0)$ for some $t \in Z^+$, so that $P(R) \subseteq Rb^t = \{Rb\}^t = \{Rc + P(R)\}^t \subseteq Rc$. We may now repeat the above argument with $b = c$ and conclude that $\ell_R(c) = (0)$. This completes the proof.

4.7. Proofs of theorems 4.2 and 4.3. Suppose R is a fully left Goldie ipli-ring such that $R/P(R)$ is a prime ring. Let Ra and Rb be ideals of R such that Rb is non-nilpotent and $RaRb = (0)$. By 4.4, we have $RaRb = Rab$; so $ab = 0$. 4.6 now shows that $a = 0$ so that $Ra = (0)$. Thus, R is a primary ring.

Suppose R is a primary fully left Goldie ipli-ring. Assume for a moment that $R/P(R)$ is not a prime ring. Let $\bar{A}$, $\bar{B}$ be non-zero ideals of $\bar{R} = R/P(R)$ such that $\bar{A}\,\bar{B} = (\bar{0})$. Let A, B be the full inverse images of $\bar{A}$ and $\bar{B}$ under the canonical epimorphism $R \to R/P(R)$. Then $AB \subseteq P(R)$; by 4.5, AB is nilpotent. Let t be the index of nilpotency of AB. Since R is primary and A, B are both non-nilpotent, we have $t > 1$. Now, $(0) = (AB)^t = \{(AB)^{t-1}A\}\,B$ implies $(0) = \{(AB)^{t-1}\}\,A$ which, in turn, gives $0 = (AB)^{t-1}$. However, this is contrary to our choice of t. Hence $R/P(R)$ is a prime ring. This proves Johnson's second theorem. Johnson's third theorem now follows from lemma 4.6.

4.8. COROLLARY. Let R be a fully left Goldie ipli-ring and A be an ideal of R contained in $P(R)$. R is a primary ring if and only if R/A is a primary ring.

Proof. Immediate from Johnson's second theorem.

We now turn to Johnson's first theorem. The following lemma contains the semi-prime case of the theorem.

4.9. LEMMA. Let R be a semi-prime left Goldie ipli-ring. Then R is a direct sum of a finite number of ideals, each of which is a prime left Goldie ipli-ring.

Proof. By Goldie's second theorem, R has a semi-simple Artinian total l.q. ring, say Q. There exist mutually orthogonal non-zero central idempotents $e_1, ----, e_n$ in Q, such that each e_i is primitive in the centre of Q and $\sum_{i=1}^{n} e_i = 1$. Let $S = \Sigma Re_i$. Clearly, S is a subring of Q and $R \subseteq S$. Let

$$E = \{r \in R | e_i r \in R, 1 \leq i \leq n\}.$$

Evidently, SE is an ideal of R; so $SE = Ra$ for some $a \in R$. There exist $r_i \in R$, $1 \leq i \leq n$, and a regular element c in R such that $e_i = c^{-1} r_i$ for $1 \leq i \leq n$. Since e_i are central in Q, we have

$$r_i = ce_i = e_i c;$$

so, $c \in E$. It follows that $c \in Ra$. Since c is regular in R, it must be a unit in Q. Thus, a is a unit in Q and so regular in S. Now, $Ra = SE = S \cdot SE = SRa = Sa$ shows that, for every $s \in S$, there exists $r \in R$ such that $sa = ra$ i.e., $(s-r)\,a = 0$; regularity of a in S gives $r = s$. Thus $S = R$. The lemma is now clear.

4.10. LEMMA. Let R be a fully left Goldie ipli-ring. Let $\sigma\colon R \to \overline{R} = R/P(R)$ be the canonical epimorphism. If $\overline{R} = \overline{A} \oplus \overline{B}$, where $\overline{A}$, $\overline{B}$ are non-zero ideals of $\overline{R}$, then there exist ideals A', B' of R such that $R = A' \oplus B'$, where $A' + P(R) = \sigma^{-1}(\overline{A})$ and $B' + P(R) = \sigma^{-1}(\overline{B})$.

Proof. Put $A = \sigma^{-1}(\overline{A})$ and $B = \sigma^{-1}(\overline{B})$. Then $R = A + B$. Assume that $R = A^t + B^t$, $t \in Z^+$. Then $R = (A+B)(A^t+B^t) = A^{t+1} + AB^t + BA^t + B^{t+1} \subseteq A^{t+1} + B^{t+1} + P(R)$, since AB and BA are contained in $P(R)$. However, $P(R) = RP(R) = A^t P(R) + B^t P(R) \subseteq A^{t+1} + B^{t+1}$.

Hence $R = A^{t+1} + B^{t+1}$. This completes the induction on t and shows that $R = A^n + B^n$ for every $n \in Z^+$.

It is easily seen that $A^n + P(R) = A$ and $B^n + P(R) = B$ for every $n \in Z^+$. To finish the proof, it suffices to show that $A^s \cap B^s = (0)$ for some $s \in Z^+$. Put $A = Ra$ and $B = Rb$. Clearly, $a + P(R)$ is a regular element in the ring $\bar{A}$. Thus, if $x \in R$ and $xa \in P(R)$ then $x \in B$. It follows that $P(R) = Rba = BA$. Similarly, $P(R) = AB$. An induction, using $BA = AB$ shows that $(AB)^n = A^n B^n$, $n \in Z^+$. By 4.5, there exists $m \in Z^+$ such that $(0) = \{P(R)\}^m = (AB)^m = A^m B^m$; so, $A^m \subseteq \ell_R(b^m)$. By 1.2.5, we get $s \in Z^+$ such that

$$A^s \cap Rb^s = (0) = A^s \cap B^s.$$

This completes the proof.

4.11. Proof of Theorem 4.1. Let R be a fully left Goldie ipli-ring. Then $R/P(R)$ is a semi-prime ipli-ring. Using 4.9 and 4.10 repeatedly, we can express R as $R = \bigoplus_{i=1}^{n} R_i$, where each R_i is a fully left Goldie ipli-ring and each $R_i/P(R_i)$ is a prime ring; by Johnson's second theorem, each R_i is a primary ring. The remaining part of the proof is clear.

We finish this section with some results on primary left Artinian ipli-rings.

4.12. THEOREM. Let Q be a primary left Artinian ring in which the prime radical $P(Q)$ is a principal left ideal of Q. Then

(1) There exists a set M of matrix units in Q such that the centralizer S of M in Q is a completely primary left Artinian pli-ring, every left ideal of S is two-sided and is of the form $\{P(S)\}^n$, $0 \leq n \leq k+1$, where $k+1$ = index of nilpotency of $P(S)$ = index of nilpotency of $P(Q)$.

(2) Q is a pli-ring. The only two-sided ideals of Q are those

of the form $\{P(Q)\}^n$, $0 \leq n \leq k+1$. Further, $P(Q) = Qw$ for any $w \in P(Q) \cap S$ but $w \notin \{P(Q)\}^2 \cap S$.

(3) If $P(Q) = Qz$ then $\ell_Q(z^k) = P(Q)$.

Proof. Since Q is a left Artinian primary ring, $Q/P(Q)$ is a simple Artinian ring. Using the Wedderburn-Artin theorem and proposition 1.2, we obtain a system M of matrix units in Q such that the centralizer S of M in Q is a completely primary left Artinian ring; so, $Q \cong M_m(S)$. The rest of the assertions in (1) and (2) are proved in Jacobson [2, pages 76-77].

We prove (3). Identify Q with $M_m(S)$. Let $P(S) = Sw_o$. Then $S\,w_0 \subseteq \ell_S\,(w_0^k)$. Since all the left ideals of S are of the form $\{P(S)\}^n = Sw_o^n$, it follows that $\ell_S(w_0^k) = Sw_o$. Now, putting $w_0^* =$ diag w_0, it is easily seen that $P(Q) = Qw_o^*$ and that $\ell_Q(w_0^{*k}) = Qw_o^* = P(Q)$. Clearly, $Qz^k = Qw_o^{*k}$. So, $z^k = aw_o^{*k}$ and $w_0^{*k} = bz^k$ for some $a, b \in Q$. Since $(1-ba)w_0^{*k} = 0$, $1-ba \in P(Q)$. It follows that a, b are units in Q. If $qz^k = 0$ then $qaw_o^{*k} = 0$ so $qa \in \ell_Q(w_0^{*k}) = P(Q)$ so $q \in P(Q)$. Hence $\ell_Q(z^k) \subseteq P(Q)$; the other inclusion is trivial. This completes the proof.

References.

Theorems similar to Johnson's first theorem are well-known and classical e.g. the structure theorem of commutative PIR's (see Zariski and Samuel [1,page 245]) and the structure theorem for left-right Artinian pli-rings (see Jacobson [2, page 75]). Goldie [3] proved a theorem similar to Johnson's first theorem for right Noetherian pli-rings (see §8 of this chapter). Johnson [1] contains the three theorems stated in the text for pli-rings. For some time, it was not clear that Johnson's theorems are genuinely stronger that Goldie's result; in fact, the following conjecture of Herstein [1, page 75] was billed as highly likely to be true: 'A primary pli-ring is either prime or Artinian'. In effect, the conjecture states that Johnson's theorems contain the same information as Goldie's fourth theorem (proved in §8). Jategaonkar [3] contains a counter-example to Herstein's conjecture; there was an earlier unpublished paper by the author, Jategaonkar [2], containing a negative solution of Jacobson's conjecture on intersection of powers of Jacobson radical and a negative solution of Herstein's conjecture.

Johnson's theorems were adopted by Robson [3] for left Noetherian ipli-rings. Our proofs are taken from Robson [3] with trivial changes.

§5. FULLY LEFT GOLDIE SEMI-PLI-IPLI-RINGS.

The following theorem is basic for the study of fully left Goldie semi-pli-ipli-rings (in particular, pli-rings).

5.1. THEOREM. Let R be a fully left Goldie semi-pli-ipli-ring. An element $c \in R$ is left regular in R if and only if $c + P(R)$ is left regular in $R/P(R)$. If c is left regular in R then $P(R) \subset Rc$.

Proof. Due to Johnson's first theorem, it suffices to prove the theorem under the additional hypothesis that R is a primary ring. Let $P = P(R)$ and let $k+1$ be the index of nilpotency of P. If $P = (0)$, there is nothing to prove. Assume that $P \neq (0)$, so $k \geq 1$.

(i) Let c be a regular element of R. If $x \in R$ and $xc \in P$ then $P^k xc = (0) = P^k x$, so $x \in \mathfrak{t}(P^k)$. Since P^k is a two-sided ideal of R, $\mathfrak{t}(P^k)$ is also a two-sided ideal of R. By Johnson's third theorem, either $\mathfrak{t}(P^k)$ is nilpotent or contains a left regular element; in the later case, we get $P^k = (0)$, contrary to our choice of k. Thus, $x \in \mathfrak{t}(P^k) \subseteq P$. Thus, $c + P$ is left regular in R/P.

(ii) Suppose $d \in R$ and $d + P$ is a left regular element in R/P. Since R is ipli, P is principally generated; so $Rd + P$ is a finitely generated left ideal of R. Since R is semi-pli, $Rd + P = Ra$ for some $a \in R$. Then $(d+P) = (r+P)(a+P)$, where $r \in R$. By Goldie's first theorem, R/P is a left order is a simple Artinian ring Q. By 1.2.10, $d+P$ is regular in R/P, so a unit in Q. Thus, $a+P$ is regular in R/P. Evidently, $a = r_1 d + w_1$ for some $r_1 \in R$ and $w_1 \in P$. Since $P \subseteq Ra$, $w_1 = wa$; regularity of $a+P$ in R/P shows that $w \in P$; so $(1-w)$ is a unit in R. Now, $(1-w)a = r_1 d$ shows that

$$Ra = Rd \supset P.$$

Since $d^n + P$ is also left regular in R/P, the above argument shows that $P \subset Rd^n$ for every $n \in Z^+$. The left regularity of $d+P$ in R/P implies $\ell(d) \subseteq P$. By 1.2.5, for sufficiently large m, we have $(0) = \ell(d) \cap Rd^m$. Hence $\ell(d) = (0)$.

(iii) Suppose c is left regular in R. Part (i) of the proof shows that $c+P$ is left regular in R/P; so part (ii) is applicable to $c = d$. Thus, $P \subset Rc$. This completes the proof.

<u>5.2. COROLLARY</u>. Let R be a fully left Goldie semi-pli-ipli-ring and let A be a nilpotent ideal of R. An element $c \in R$ is left regular in R if and only if $c+A$ is left regular in R/A.

<u>Proof</u>. Immediate from 5.1.

<u>5.3. COROLLARY</u>. Let R be a fully left Goldie semi-pli-ipli-ring. R is a left order in a left Artinian ring if and only if every left regular element in R is regular.

<u>Proof</u>. Follows from 5.1 and Small's theorem.

<u>References</u>.

The results of this section are due to the author. They may be found in Jategaonkar [1;5].

§6. ASSOCIATED GRADED RINGS.

Let R be a ring and A be a proper ideal of R. Let $A^o = R$ and let

$$G_A(R) = \bigoplus_{n\geq 0} A^n/A^{n+1}$$

as an abelian group. Define a multiplication in $G_A(R)$ by assuming the distributive laws and the rule

$$(r_m + A^{m+1})\ (s_n + A^{n+1}) = r_m\, s_n + A^{m+n+1},$$

where $r_m \in A^m$ and $s_n \in A^n$. It is straightforward to check that $G_A(R)$ is a ring. $G_A(R)$ is called the <u>graded ring associated with the filtration on R induced by powers of A</u>.

In this section, we describe $G_P(R)$, where R is a primary fully left Goldie semi-pli-ipli-ring and $P = P(R)$. Once this is done, using Johnson's first theorem, one can obtain a similar description of $G_P(R)$ for arbitrary fully left Goldie semi-pli-ipli-rings.

Let R be a primary fully left Goldie semi-pli-ipli-ring. Let

$$T(R) = \{r \in R \mid cr = 0 \text{ for some left regular } c \text{ in } R\}.$$

Thus, by 5.3, R has a left Artinian l.q. ring if and only if $T(R) = (0)$. The starting point of our discussion of $G_P(R)$ is the following theorem.

<u>6.1. THEOREM</u>. Let R be a primary fully left Goldie semi-pli-ipli-ring. Let $k+1$ be the index of nilpotency of $P(R)$. Then either $T(R) = (0)$ or $T(R) = \{P(R)\}^k$. In any case, $R/T(R)$ is a primary fully left Goldie semi-pli-ipli-ring with a left Artinian l.q. ring.

This theorem provides the following rough classification of primary fully left Goldie semi-pli-ipli-rings. Let R be such a ring. If $P(R) = (0)$, then, by Johnson's second theorem, R is a prime ring. If $P(R) = T(R) \neq (0)$, we shall say that R has <u>small prime radical</u>. If R is neither a prime ring nor a primary ring with small prime radical then R has <u>large prime radical</u>. Notice that R has large prime radical if $\{P(R)\}^2 \neq (0)$; however, there exist rings with large prime radical such that $\{P(R)\}^2 = (0)$. It may be worthwhile to note that there exist primary pli-rings with small prime radical.

We now state the main theorem of this section.

<u>6.2. THEOREM</u>. Let R be a primary fully left Goldie semi-pli-ipli-ring with large prime radical P. Let k+1 be the index of nilpotency of P, so $k \in Z^+$. Let $\bar{R} = R/P$ and let $\bar{Q}$ be the simple Artinian l.q. ring of $\bar{R}$ (which exists by Johnson's second theorem and Goldie's first theorem). Then there exists a monomorphism $\rho\colon \bar{Q} \to \bar{R}$ such that $G_P(R) \cong S/I$ where $S = \bar{R}[x, \rho|\bar{R}]$ and I is an ideal of S with $Sx^{k+1} \subseteq I \subset Sx^k$. R has a left Artinian l.q. ring if and only if $I = Sx^{k+1}$.

Assume further that R is a pli-ring. Then $G_P(R)$ is a primary pli-ring with large prime radical. R has a left Artinian l.q. ring if and only if $G_P(R)$ has a left Artinian l.q. ring. If R does not have a left Artinian l.q. ring then there exists a regular element $\bar{c}$ in $\bar{R}$ such that $I = S\bar{c}x^k$, so $G_P(R) \cong S/S\bar{c}x^k$.

Until further notice, let R be a ipli-ring, $P = P(R)$, k+1 = index of nilpotency of P and $\bar{R} = R/P$. If $r \in R$, put $\bar{r} = r+P$.

The following lemma shows that $G_P(R)$ and $\bar{R}[x,\rho]$ are related.

<u>6.3. LEMMA</u>. Assume that there exists $w_o \in P$ such that $P = Rw_o$, $\ell(w_o) \subseteq P$ and $\mathfrak{r}(w_o^k) \subseteq P$. Then there exists a monomorphism $\rho\colon \bar{R} \to \bar{R}$ and an epimorphism $\varphi\colon S \to G_P(R)$, $S = \bar{R}[x,\rho]$, such that $\varphi(x) = w_o + P^2$ and $\varphi|\bar{R} = \mathrm{id}$.

Further, if $\ell(w_o^k) = P$ then $\ker \varphi = Sx^{k+1}$ and each left $\bar{R}$-module P^n/P^{n+1}, $n \leq k$, is $\bar{R}$-isomorphic with ${}_{\bar{R}}\bar{R}$.

Proof. Since Rw_o is a two-sided ideal of R, for every $r\epsilon R$, there exists $r'\epsilon R$ such that $w_0 r = r'w_0$. Since $\ell(w_0) \subseteq P$, $\overline{r}'$ is uniquely determined by r. If $r\epsilon P$, say $r = sw_o$, then $w_0 r = (s'w_0)w_0$; so $\overline{r}' = \overline{0}$. It follows that $\overline{r} \mapsto \overline{r}'$ defines a mapping $\rho\colon\ \overline{R} \to \overline{R}$. If $\rho(\overline{r}) = \overline{0}$ then $w_0 r \in P^2$; so, $w_0^k r = 0$ and $r \in \ell(w_0^k) \subseteq P$; i.e. $\overline{r} = \overline{0}$. It is now clear that ρ is a monomorphism.

Observe that, in $G_P(R)$, $w_0^n + P^{n+1} = (w_0 + P^2)^n$ for every $n\epsilon Z^+$. Thus, every element of $G_P(R)$ can be uniquely expressed as

$$\sum_{n=0}^{k} \overline{r}_n\ (w_0 + P^2)^n,$$

where $\overline{r}_n \in \overline{R}$. Define a mapping $\varphi\colon\ S \to G_P(R)$ by

$$\varphi(\sum_{n\geq 0} \overline{r}_n\ x^n) \ = \ \sum_{n\geq 0} \overline{r}_n\ (w_0 + P^2)^n.$$

Clearly, φ is an epimorphism of left $\overline{R}$-modules. Also, $\varphi(fx) = \varphi(f)\varphi(x)$ for every $f \in S$. By an induction, we have $\varphi(x^n\overline{r}) = \varphi(x^n)\varphi(\overline{r})$ for every $n \in Z^+$ and $\overline{r} \in \overline{R}$. Now, let $f, g \in S$. Express g as $g = \overline{r} + g_1 x$, where $\overline{r} \in \overline{R}$ and $g_1 \in S$. Then $\varphi(fg) = \varphi(f\overline{r} + fg_1 x) = \varphi(f)\varphi(\overline{r}) + \varphi(fg_1)\varphi(x)$. Thus an induction on the degree of g is available to conclude that $\varphi(fg) = \varphi(f)\varphi(g)$. Hence φ is an epimorphism of rings with $\varphi(x) = w_0 + P^2$ and $\varphi|\overline{R} = \text{id}$.

If $\sum_{n=0}^{k} \overline{r}_n\ x^n \in \ker\varphi$ then $\sum_{n=0}^{k} \overline{r}_n(w_0 + P^2)^n = \overline{0}$ so that $\overline{0} = \overline{r}_n(w_0 + P^2)^n = r_n\ w_0^n + P^{n+1}$; i.e., $r_n\ w_0^n \in P^{n+1}$ for every $n \leq k$. Thus $r_n\ w_0^k = 0$, $0 \leq n \leq k$. Now, if $\ell(w_0^k) = P$, then each $\overline{r}_n = \overline{0}$ and $\ker\varphi = Sx^{k+1}$. The map

$$\overline{r} \mapsto \overline{r}\ (w_0^n + P^{n+1})$$

is evidently an isomorphism of left $\overline{R}$-modules for $0 \leq n \leq k$. This proves the lemma.

6.4. LEMMA. Let R be a primary fully left Goldie ipli-ring.
(i) If $P = Rw_o$ then $\ell(w_0^k) = P$.

(ii) Assume that $P = Rw_o$ where $\ell(w_0) \subseteq P$. Then lemma 6.3 is applicable. Let $\rho\colon \bar{R} \to \bar{R}$ be the monomorphism defined in 6.3. If $c \in R$, $P \subseteq Rc$ and $\bar{c}$ is left regular in $\bar{R}$ then $\rho(\bar{c})$ is a unit in $\bar{R}$. In particular, if $\bar{R}\,\bar{r}$ is a non-zero ideal of $\bar{R}$ then $\rho(\bar{r})$ is a unit in $\bar{R}$.

(iii) Assume further that R is a semi-pli-ring. Then $\rho(\bar{d})$ is a unit in $\bar{R}$ for every left regular $\bar{d}$ in $\bar{R}$ and $\rho\colon \bar{R} \to \bar{R}$ can be uniquely extended to a monomorphism $\rho\colon \bar{Q} \to \bar{R}$, where $\bar{Q}$ is the simple Artinian l.q. ring of $\bar{R}$.

<u>Proof</u>. (i) If $P = Rw_o$ then $t(w_0^k) = t(P^k)$ is clearly an ideal of R and contains P. Johnson's third theorem implies that $t(w_0^k) \subseteq P$; so $t(w_0^k) = P$.

(ii) Assume that $Rw_o = P$ with $\ell(w_0) \subseteq P$. Let $c \in R$, $P \subseteq Rc$ and let $\bar{c}$ be left regular in $\bar{R}$. Evidently, we have $a \in R$ with $w_0 = ac$ so $\bar{0} = \bar{a}\,\bar{c}$; left regularity of $\bar{c}$ implies $a \in Rw_o$, say $a = bw_o$. Then $\bar{b}\rho(\bar{c}) = 1$. By Goldie's first theorem, $\bar{R}$ has a simple Artinian l.q. ring $\bar{Q}$. In $\bar{Q}$, $\bar{b}\rho(\bar{c}) = \bar{1}$ implies $\rho(\bar{c})\bar{b} = \bar{1}$. It follows that $\rho(\bar{c})$ is a unit in $\bar{R}$.

If $\bar{R}\,\bar{r}$ is a non-zero ideal of $\bar{R}$, pick $s \in R$ such that $Rr + P = Rs$. By Johnson's third theorem $\bar{r} = \bar{s}$ is left regular in $\bar{R}$. The above argument now shows that $\rho(\bar{r})$ is a unit in $\bar{R}$.

(iii) Besides assumptions in (ii), if R is assumed to be a semi-pli-ring, then theorem 5.1 becomes available. It is now easy to complete the proof.

It remains to discover rings to which 6.3 and 6.4 are applicable. Observe that in both these lemmas, the main hypothesis is the existence of a generator for P with 'small' annihilators.

6.5. LEMMA. Let R be a primary fully left Goldie ipli-ring with a non-zero prime radical P and with a left Artinian l.q. ring Q. Then for every $w_0 \in R$ such that $P = Rw_0$, we have $\ell_R(w_0^k) = P$.

Proof. Let A be a left ideal of Q. The common denominator theorem yields $A = Q(A \cap R)$. It follows that Q is a ipli-ring. By 1.4.5, $Q/P(Q)$ is a total l.q. ring of $\bar{R}$; so, $Q/P(Q)$ is a simple Artinian ring. By Johnson's second theorem, Q is a primary ipli-ring. Let $P = Rw_0$. By 1.4.5, $P(Q) = QP(R) = Qw_0$ and $P(R) = P(Q) \cap R$. Thus the index of nilpotency of $P(Q)$ is $k+1$, the index of nilpotency of $P(R)$. By theorem 4.12, $\ell_Q(w_0^k) = Qw_0$. So, $\ell_R(w_0^k) = \ell_Q(w_0^k) \cap R = Qw_0 \cap R = P(Q) \cap R = P(R)$. This proves the lemma.

6.6. PROPOSITION. Let R be a primary fully left Goldie ipli-ring with a non-zero prime radical and with a left Artinian l.q. ring. Then there exists a monomorphism $\rho\colon \bar{R} \to \bar{R}$ such that

$$G_P(R) \cong S/Sx^{k+1},$$

where $S = \bar{R}[x,\rho]$.

Let $F = \{\bar{r} \in \bar{R} \mid \bar{r} \neq 0,\ \bar{R}\bar{r} \text{ an ideal of } \bar{R}\}$. Then F is a left divisor set in $\bar{R}$. Let L be the l.q. ring of $\bar{R}$ w.r.t. F. Then L is a simple left Goldie ring and ρ can be uniquely extended to a monomorphism $\rho\colon L \to \bar{R}$.

Proof. The first part of the proposition follows from 6.3, 6.4 and 6.5.

Using 4.4, it is easily seen that F is a left divisor set in $\bar{R}$. Since $\bar{R}$ has a simple Artinian l.q. ring $\bar{Q}$ with $\bar{R} \subseteq L \subseteq \bar{Q}$, 1.2.14 and Goldie's first theorem show that L is a prime left Goldie ring. Since $A = L(A \cap R)$ holds for every left ideal A of R, it follows that L is a simple ring. By 6.4, $\rho(\bar{r})$ is a unit in $\bar{R}$ for every $\bar{r} \in F$; so ρ can be uniquely extended to a monomorphism $\rho\colon L \to \bar{R}$. This completes the proof.

6.7. LEMMA. Let R be a primary fully left Goldie semi-pli-ipli-ring. Then $T(R)$ is a nilpotent two-sided ideal of R and $P(R)T(R) = (0)$. Further, $R/T(R)$ has a left Artinian l.q. ring.

Proof. If $T(R) = (0)$ then the lemma follows from 5.3. Assume that $T(R) \neq (0)$. Let $x_1, x_2 \in T(R)$ and c_1, c_2 be left regular elements in R such that $c_1 x_1 = 0 = c_2 x_2$. Since $\bar{R}$ is a prime left Goldie ring, it has the left common multiple property; by 5.1, $\bar{c}_1$ and $\bar{c}_2$ are left regular in $\bar{R}$, so regular in $\bar{R}$ by 1.2.10; thus, there exist $d_1, d_2, d \in R$ such that $\bar{d}_1, \bar{d}_2, \bar{d}$ are regular in $\bar{R}$ and $\bar{d}_1 \bar{c}_1 = \bar{d}_2 \bar{c}_2 = \bar{d}$. So, $d = d_1 c_1 + w_1 = d_2 c_2 + w_2$ for some $w_1, w_2 \in P(R)$. By 5.1, d is left regular in R and $P \subseteq R c_1$, $P \subseteq R c_2$. Let $w_1 = r_1 c_1$; $w_2 = r_2 c_2$. Then $d = (d_1 + r_1) c_1 = (d_2 + r_2) c_2$. It follows that $d(x_1 - x_2) = 0$, so $x_1 - x_2 \in T(R)$.

Let $x \in T(R)$ and $bx = 0$ where b is left regular in R. Let $r \in R$. An argument similar to the above one shows that $b_1 r = r_1 b$, where $b_1, r_1 \in R$ with b_1 left regular in R. Thus $b_1 rx = 0$ and $rx \in T(R)$. It is now clear that $T(R)$ is an ideal of R.

Let $x \in T(R)$. Then $\ell(x)$ contains a left regular element of R; by 5.1, $P(R) \subseteq \ell(x)$ so $P(R)x = (0)$. Hence $P(R)T(R) = (0)$. If $T(R)$ is non-nilpotent, by Johnson's third theorem, it will contain a left regular element, contradicting the definition of $T(R)$. Thus $T(R)$ is nilpotent.

If $a + T(R)$ is left regular in $R/T(R)$ and if $(a+T(R))(y+T(R)) = 0 + T(R)$, then by 5.2, a is left regular in R and $ay \in T(R)$. So, there exists a left regular $f \in R$ with $fay = 0$. Since fa is left regular in R, $y \in T(R)$. 5.3 now shows that $R/T(R)$ has a left Artinian l.q. ring. This completes the proof.

6.8. Proof of Theorem 6.1. We put an induction on the index of nilpotency, $k+1$, of $P(R)$ to show that either $T(R) = (0)$ or $T(R) = \{P(R)\}^k$.

If $k = 0$, then R is a prime left Goldie ring; I.2.10 shows that $T(R) = (0)$.

Let $k > 0$. Assume that $T(R) \neq (0)$. Also assume that the lemma holds for all primary fully left Goldie semi-pli-ipli-rings for which the index of nilpotency of the prime radical is less than $k+1$. In order to complete the induction, we now have to show that $T(R) = \{P(R)\}^k$.

By 6.7, $T(R) \subseteq P(R)$. Since $P(R)$ is nilpotent and $T(R)$ is assumed to be non-zero, there exists $m \in Z^+$ such that $T(R) \subseteq \{P(R)\}^m$ and $T(R) \not\subseteq \{P(R)\}^{m+1}$. Suppose $m < k$; we proceed to obtain a contradiction.

Let σ: $R \to S = R/\{P(R)\}^{m+1}$ be the canonical epimorphism. Using Johnson's second theorem, it follows that S is a primary fully left Goldie semi-pli-ipli-ring. Evidently, $P(S) = \sigma(P(R)) = P(R)/\{P(R)\}^{m+1}$; so, the index of nilpotency of $P(S)$ is $m+1 < k+1$. By our induction hypothesis, either $T(S) = (\sigma(0))$ or $T(S) = \{P(S)\}^m$. Pick $x \in T(R)$, $x \notin \{P(R)\}^{m+1}$. Let c be a left regular element in R with $cx = 0$ so that $\sigma(c)\sigma(x) = \sigma(0)$. By 5.2, $\sigma(c)$ is left regular in S; $\sigma(x) \neq \sigma(0)$ by our choice. So, $T(S) \neq (\sigma(0))$. We thus have $T(S) = \{P(S)\}^m$.

Let $P(R) = Rw_0$. Then $\{P(R)\}^n = Rw_0^n$ and $\{P(S)\}^n = S\{\sigma(w_0)\}^n$ for every $n \in Z^+$. Consequently, $\sigma(w_0^m) \in \{P(S)\}^m = T(S)$. Choose $d \in R$ such that $\sigma(d)$ is left regular in S and $\sigma(d)\sigma(w_0^m) = \sigma(0)$. Then $dw_0^m = r_1 w_0^{m+1}$; i.e., $(d-r_1 w_0)w_0^m = 0$ for some $r_1 \in R$. A repeated use of 5.2 shows that $d - r_1w_0$ is left regular in R. Thus $w_0^m \in T(R)$ and $\{P(R)\}^m \subseteq T(R)$. Our choice of m now yields $\{P(R)\}^m = T(R)$. By 6.7, $(0) = P(R)T(R) = \{P(R)\}^{m+1}$, a contradiction. Hence $m = k$ and $T(R) \subseteq \{P(R)\}^k$.

Assume that $(0) \neq T(R) \subsetneq \{P(R)\}^k$. We proceed to obtain a contradiction. Consider $\{P(R)\}^k$ canonically as a left module over the ring $\overline{R} = R/P(R)$. Then $T(R)$ is a non-zero proper left $\overline{R}$-submodule of the cyclic left $\bar{R}$-module $\{P(R)\}^k$. Hence ${}_{\bar{R}}\{P(R)\}^k/{}_{\bar{R}}T(R)$ is a non-zero proper factor module of the left $\bar{R}$-module ${}_{\bar{R}}\bar{R}$.

As shown in 6.7, $T(R)$ is an ideal contained in $P(R)$. Consider the ring $A = R/T(R)$. Johnson's second theorem and lemma 6.7 show that A is a primary fully left Goldie semi-pli-ipli-ring with a left Artinian l.q. ring. Since we have assumed that $T(R) \subsetneqq \{P(R)\}^k$, the index of nilpotency of $P(A)$ is k+1. 6.3, 6.4 and 6.5 show that if we consider $\{P(A)\}^k$ canonically as a left module over the ring $\bar{A} = A/P(A)$ then it is $\bar{A}$-isomorphic with ${}_{\bar{A}}\bar{A}$. Let ψ: $\bar{A} \to \bar{R}$ be the isomorphism defined by $(r + T(R)) + P(A) \mapsto r + P(R)$. Consider the mapping

$$\varphi: \ {}_{\bar{A}}\{P(A)\}^k \to {}_{\bar{R}}\{P(R)\}^k/{}_{\bar{R}}T(R)$$

defined as follows: Any element of $\{P(A)\}^k$ is of the form $w + T(R)$ for some $w \in \{P(R)\}^k$; to get $\varphi(w + T(R))$, consider w as an element of ${}_{\bar{R}}\{P(R)\}^k$ and take is image in the factor module of ${}_{\bar{R}}\{P(R)\}^k$ by ${}_{\bar{R}}T(R)$. It is straightforward to check that φ is a ψ-linear isomorphism of left modules. It follows that

$${}_{\bar{R}}\{P(R)\}^k \ / \ {}_{\bar{R}}T(R) \cong {}_{\bar{R}}\bar{R}.$$

To sum up, we have an $\bar{R}$-epimorphism u: ${}_{\bar{R}}\bar{R} \to {}_{\bar{R}}\bar{R}$ with $\ker u \neq (0)$. Clearly, there exists $\bar{a} \in \bar{R}$ such that $u(\bar{r}) = \overline{ra}$ for every $\bar{r} \in \bar{R}$. Since u is an epimorphism, $\bar{1} = \bar{b}\,\bar{a}$ for some $\bar{b} \in \bar{R}$. Since $\bar{R}$ is a left order in a simple Artinian ring, $\bar{a}$ must be a unit in $\bar{R}$, so $\ker u = (0)$, a contradiction. Hence $T(R) = \{P(R)\}^k$. This completes the induction.

The remaining assertion is contained in 6.7. This completes the proof.

<u>6.9. Proof of theorem 6.2.</u> Let R be as stated. Choose $w_0 \in R$ such that $Rw_o = P$. By 6.4, $\ell(w_0^k) = P$. We claim that $\ell(w_0) \subseteq P$. If $k = 1$, then by 6.1, $T(R) = (0)$ or $T(R) = P(R)$; since R is assumed

to have large prime radical, we have $T(R) = (0)$. By 5.3 and 6.5, we have $\ell(w_o) = P(R)$. Suppose $k > 1$. If $T(R) = (0)$ then by 5.3 and 6.5, $\ell(w_o^k) = P(R)$ so $\ell(w_o) \subseteq P(R)$. If $T(R) \neq (0)$, then by theorem 6.1, $T(R) = R\,w_o^k$ and $R/T(R)$ has a left Artinian l.q. ring. Since $P(R/T(R)) = P(R)/\{P(R)\}^k = (R/\{P(R)\}^k)\,(w_o + \{P(R)\}^k)$, using 6.5, we have

$$\ell_{R/T(R)}(w_o + \{P(R)\}^k) \subseteq P(R)/\{P(R)\}^k.$$

It follows that $\ell(w_o) \subseteq P(R)$. This proves our claim. Lemmas 6.3 and 6.4 are thus applicable to the ring R.

Let $\varphi\colon S \to G_P(R)$ and $\rho\colon \overline{Q} \to \overline{R}$ be as in lemmas 6.3 and 6.4. If $\varphi(\overline{r}_n x^n) = 0$ then $r_n w_o^n \in P^{n+1}$. If $T(R) = (0)$, then $r_n w_o^k = 0$ gives $r_n \in \ell(w_o^k) = P(R)$ so $\overline{r}_n = 0$; hence in this case, $\ker \varphi = Sx^{k+1}$. If $T(R) \neq (0)$ and $n < k$ then $r_n w_o^{k-1} \in \{P(R)\}^k = T(R)$. Since $T(R/T(R)) = (0)$, it follows that $r_n \in P(R)$ so $\overline{r}_n = 0$. Thus, in this case, $Sx^{k+1} \subset \ker \varphi \subset Sx^k$.

Now assume that R is a primary pli-ring with large prime radical. Then theorem 3.1 shows that S is a pli-ring; so $G_P(R)$ is a pli-ring. Using Johnson's second theorem, it follows that $G_P(R)$ is primary. The rest is easy. This completes the proof.

We now abandon the restrictions on R placed before lemma 6.3 and finish this section with three propositions of some interest.

6.10. PROPOSITION. Let R be a primary ipli-ring, P be a non-nilpotent proper prime ideal of R and $\overline{R} = R/P$. Then there exists a monomorphism $\rho\colon \overline{R} \to \overline{R}$ such that

$$G_P(R) \cong \overline{R}\,[x, \rho].$$

Proof: Adopt the first part of the proof of 6.3.

6.11. PROPOSITION. Let R be a primary fully left Goldie semi-pli-ipli-ring. Let $k + 1$ be the index of nilpotency of $P(R)$, where $k \in Z^+$. Then $R/\{P(R)\}^n$ has a left Artinian ℓ.q. ring for $1 \leq n \leq k$.

Proof. If R does not have a left Artinian ℓ.q. ring then by 6.1, $T(R) = \{P(R)\}^k$ and $R/\{P(R)\}^k$ has a left Artinian ℓ.q. ring. Since

$$R/\{P(R)\}^n \cong (R/\{P(R)\}^k)/\{P(R/\{P(R)\}^k)\}^n$$

for $1 \leq n \leq k$, we may assume without loss of generality that R has a left Artinian ℓ.q. ring, say Q. Using 1.4.5 and 4.12, it follows that Q is a primary pli-ring.

We now claim that $\{P(R)\}^n = \{P(Q)\}^n \cap R$ for $1 \leq n \leq k + 1$. By 1.4.5, our claim is true for $n = 1$ and $n = k + 1$. Let $1 < n < k+1$ and assume that $\{P(R)\}^{n-1} = \{P(Q)\}^{n-1} \cap R$. Let $P(R) = R\,w_0$ so that, by 1.4.5, $P(Q) = Q\,w_0$. As in 4.12, we get $\ell_Q(w_0) = Qw_0^k$. Let $r \in \{P(Q)\}^n \cap R = Q\,w_0^n \cap R$. Then $r = c^{-1}r'w_0^n$, i.e., $cr = r'w_0^n$ for some $r, c \in R$, c regular in R. By 5.1 and 1.2.10, $c + P(R)$ is regular in $R/P(R)$. Thus, $r \in P(R)$; say, $r = r_1w_0$, $r_1 \in R$. Then

$$(cr_1 - r'w_0^{n-1})w_0 = 0$$

so, $cr_1 - r'w_0^{n-1} \in \ell_Q(w_0) \cap R = Qw_0^k \cap R$. So, $cr_1 - r'w_0^{n-1} = d^{-1}r''w_0^k$, i.e.,

$$r_1 = c^{-1}d^{-1}(dr' + r''w_0^{k-n+1})w_0^{n-1}$$

for some r'', $d \in R$. Using the induction hypothesis, we have $r_1 \in \{P(Q)\}^{n-1} \cap R = \{P(R)\}^{n-1}$; so $r = r_1w_0 \in \{P(R)\}^n$. Hence $\{P(R)\}^n \supset \{P(Q)\}^n \cap R$. Since $P(Q) = QP(R)$, the other inclusion is trivial. This completes the induction and proves our claim.

We now show that $R/\{P(R)\}^n$ satisfies the regularity condition for $1 \leq n \leq k + 1$. Let b be a regular element of R. If $x \in R$ and $xb \in Rw_0^n$ then $x \in Rw_0^nQ \cap R \subseteq Qw_0^n \cap R = Rw_0^n$. Similarly, if $y \in R$ and $by \in Rw_0^n$ then $y \in Rw_0^n$. Hence $b + \{P(R)\}^n$ is regular in $R/\{P(R)\}^n$. By Small's theorem, R has the regularity condition. It follows that $R/\{P(R)\}^n$ also has the regularity condition. By Small's theorem, $R/\{P(R)\}^n$ has a left Artinian ℓ.q. ring. This completes the proof.

We need some definitions to state the next proposition in a compact form; these are due to Levy [1].

Let ${}_RM$ be a left R-module. An element $x \in M$ is a torsion element if $ax = 0$ for some regular element a in R. ${}_RM$ is a torsion module if every element of M is a torsion element. ${}_RM$ is a torsion-free module if 0 is the only torsion element of M. ${}_RM$ is a divisible module if $aM = M$ for every regular element a in R.

6.12. PROPOSITION. Let R be a primary fully left Goldie semi-pli-ipli-ring with $T(R) \neq (0)$. Then $T(R)$ can be canonically considered as a bimodule over the ring $\overline{R} = R/P(R)$. If so considered, it has the following properties.

(i) $T(R)_{\overline{R}}$ is faithful, torsion-free and divisible.

(ii) ${}_{\overline{R}}T(R)$ is cyclic and torsion.

Proof. By theorem 6.1, $T(R) = \{P(R)\}^k$, where $k + 1$ is the index of nilpotency of $P(R)$. The proposition now follows by using 5.1 and Johnson's third theorem.

Notice that, in effect, this proposition determines the structure of $G_P(R)$ when P is small. cf. Jacobson [3, page 25]. We omit details. Also the converse of 3.1 is now easy.

References.

The results of this section are due to the author. They may be found in Jategaonkar [1,5]. The author is indebted to Professor R. E. Johnson for pointing out various simple and elegant proofs.

§7. ASSOCIATED GRADED RINGS (CONTINUED).

In the last section, we have shown that if R is a primary pli-ring with large prime radical then so also is $G_{P(R)}(R)$. We now consider the converse question viz., if R is a ring such that $G_{P(R)}(R)$ is a primary pli-ring with large prime radical, is it true that R is a primary pli-ring with large prime radical? In its full generality, we have been unable to answer this question. However, we shall show in this section that if $\bar{R} \cong G_{P(R)}(R)/P(G_{P(R)}(R))$ is assumed to be a full matrix ring over a pli-domain then the above question has an affirmative answer; we also consider the structure of such rings.

The main results of this section are the following.

7.1. THEOREM. Let R be a primary pli-ring with large prime radical. Then the following conditions on R are equivalent:

(1) $R/P(R) \cong M_n(D)$, where D is a pli-domain.

(2) $R \cong M_n(S)$, where S is a completely primary pli-ring.

If R satisfies the above equivalent conditions then n and S are uniquely determined by R upto isomorphism.

7.2. THEOREM. Let R be an arbitrary ring. Assume that $G_{P(R)}(R)$ is a primary pli-ring with large prime radical and that $G_{P(R)}(R)/P(G_{P(R)}(R)) \cong M_n(D)$ where D is a pli-domain. Then R is a primary pli-ring with large prime radical. In fact, $R \cong M_n(S)$, where S is a completely primary pli-ring with large prime radical.

We need some auxiliary results before we can prove theorem 7.1 and 7.2.

7.3. PROPOSITION. Let R be a primary fully left Goldie semi-pli-ipli-ring with large prime radical. Then there exists a positive integer n and a completely primary ring S such that $R \cong M_n(S)$.

S can be chosen to satisfy the following conditions:

(a) $P(S)$ is nilpotent and $P(S) = Sz_0$ for some $z_0 \in S$;

(b) every non-nilpotent element of S is left regular in S;

(c) If I is a left ideal of S such that $I \subseteq \{P(S)\}^m$ but $I \nsubseteq \{P(S)\}^{m+1}$ then $\{P(S)\}^{m+1} \subset I$.

Proof. Put $\overline{R} = R/P(R)$ and $k+1$ = index of nilpotency of $P(R)$. By Johnson's second theorem and Goldie's first theorem, $\overline{R}$ is a left order in a simple Artinian ring $\overline{Q}$. Let $P(R) = Rw_0$. By 6.3 and 6.4, there exists a monomorphism ρ: $\overline{Q} \to \overline{R}$ such that, for every $\overline{r} \in \overline{R}$, we have

$$(w_0 + \{P(R)\}^2)\cdot\overline{r} = \rho(\overline{r})\ (w_0 + \{P(R)\}^2)$$

in the $\overline{R}$-bimodule $P(R)/\{P(R)\}^2$. Choose a complete system of matrix units $\{\overline{e}_{ij}: 1 \leq i, j \leq n\}$ in the simple Artinian ring $\rho(\overline{Q})$. Then $\{\rho(\overline{e}_{ij}): 1 \leq i, j \leq n\}$ is also a complete system of matrix units in $\rho(\overline{Q})$. Thus, (cf. Jacobson [3, page 59]), there exists a unit $\overline{u}$ in $\rho(\overline{Q})$ such that

$$\overline{u}\ \rho(\overline{e}_{ij})\ \overline{u}^{-1} = \overline{e}_{ij} \quad ; i, j = 1, ---, n.$$

Let $u \in R$ be a lift of $\overline{u}$. It is easily seen that u is a unit in R. Put $v_0 = uw_0$, so $P(R) = Rv_0$. Also, in the $\overline{R}$-bimodule $P(R)/\{P(R)\}^2$, we have

$$(v_0 + \{P(R)\}^2)\ \overline{e}_{ij} = \overline{e}_{ij}\ (v_0 + \{P(R)\}^2).$$

Since $P(R)$ is nilpotent, we can lift $\{\overline{e}_{ij}: 1 \leq i, j \leq n\}$ to a system of matrix units $\{e_{ij}: 1 \leq i, j \leq n\}$ in R. cf. Jacobson [3, page 55]. Clearly, $v_0\, e_{ij} - e_{ij}\, v_0 \in \{P(R)\}^2$ for $1 \leq i, j \leq n$.

Let S be the centralizer of $\{e_{ij}: 1 \leq i, j \leq n\}$ in R. Then $r \mapsto (s_{ij})$, $s_{ij} = \sum_{k=1}^{n} e_{ki}\, r\, e_{jk}$, defines an isomorphism φ: $R \to M_n(S)$; Jacobson [3, page 52]. Since $P(M_n(S)) = M_n(P(S))$, φ induces an isomorphism

$\overline{\varphi}$: $\overline{R} \to M_n(S/P(S))$. Using theorem 2.1, it follows that $S/P(S)$ is a domain; so, S is a completely primary ring.

Let s be a non-nilpotent element of S. If $s_1 \in S$ and $s_1 s \in P(S)$ then $s_1 \in P(S)$, since $S/P(S)$ is a domain. Consider the element diag $s \in M_n(S)$. If $(s_{ij})\text{diag } s \in P(M_n(S)) = M_n(P(S))$ then, as seen above, each $s_{ij} \in P(S)$ so that $(s_{ij}) \in P(M_n(S))$. Hence diag $s + P(M_n(S))$ is a left regular element in $M_n(S)/P(M_n(S))$. By theorem 5.1, diag s is left regular in $M_n(S)$ and $P(M_n(S)) \subseteq M_n(S)$ diag s. It follows that s is left regular in S and $P(S) \subseteq Ss$.

Consider the ring $A = R/\{P(R)\}^2$. Clearly, $P(A) = A\,(v_0 + \{P(R)\}^2)$; also, $\{e_{ij} + \{P(R)\}^2 \colon 1 \leq i, j \leq n\}$ is a system of matrix units in A with

$$(v_0 + \{P(R)\}^2)(e_{ij} + \{P(R)\}^2) = (e_{ij} + \{P(R)\}^2)(v_0 + \{P(R)\}^2).$$

Hence $v_0 + \{P(R)\}^2$ is in the centralizer of the system of matrix units $\{e_{ij} + \{P(R)\}^2\colon 1 \leq i, j \leq n\}$ in A. When transfered to $M_n(S)$ by φ, this shows that $M_n(S)\,\varphi\,(v_0) = P(M_n(S))$ and $\varphi(v_0)$ is a scalar matrix modulo entries from $\{P(S)\}^2$. Let $z_0 \in S$ such that $\varphi(v_0) - \text{diag } z_0 \in \{P(M_n(S))\}^2$. It follows that for every $z \in P(S)$, there exist $s \in S$ and $z^* \in \{P(S)\}^2$ such that

$$z = s\,z_0 + z^*.$$

Clearly, $\{P(S)\}^{k+1} = (0) \subseteq Sz_0$. Suppose $\{P(S)\}^{m+1} \subseteq Sz_0$ and $z \in \{P(S)\}^m$. Then

$$z = \sum_i z_{1i} \text{ ---- } z_{mi},$$

where $z_{ji} \in P(S)$. As observed above,

$$z_{mi} = s_i\,z_0 + z_{mi}^*$$

where $s_i \in S$ and $z_{mi}^* \in \{P(S)\}^2$. It follows that $z \in Sz_0$ so

$\{P(S)\}^m \subseteq Sz_0$. Repeating this argument, we obtain $P(S) = Sz_0$. By 4.4, $\{P(S)\}^m = Sz_0^m$ for every $m \in Z^+$.

Let I be a left ideal of S such that $I \subseteq \{P(S)\}^m$ but $I \nsubseteq \{P(S)\}^{m+1}$ for some $m \geq 0$. If $m = 0$, I contains a non-nilpotent element s of S and as shown above, $P(S) \subseteq Ss \subseteq I$. Assume that $m > 0$. There exists $s' \in S$, $s' \notin P(S)$ such that $s' z_0^m \in I$ but $s z_0^m \notin \{P(S)\}^{m+1}$. Since $Ss \supseteq P(S)$, there exists $s'' \in S$ with $z_0 = s''s'$. So, $z_0^{m+1} = s''s'z_0^m \in I$. Consequently, $\{P(S)\}^{m+1} \subseteq I$. This completes the proof.

7.4. PROPOSITION. Let S be a completely primary ring. S is a pli-ring if and only if S satisfies the following conditions:

(a) $P(S)$ is nilpotent and $P(S) = Sz_0$ for some $z_0 \in S$.

(b) $S/P(S)$ is a pli-domain.

(c) If I is a left ideal of S such that $I \subseteq \{P(S)\}^m$, $I \nsubseteq \{P(S)\}^{m+1}$ then $\{P(S)\}^{m+1} \subseteq I$.

Proof. Let S be a completely primary ring satisfying the conditions (a), (b) and (c). By 4.4, $\{P(S)\}^m = Sz_0^m$ for every $m \geq 0$. Let I be a non-zero left ideal of S. Since $P(S)$ is nilpotent, there exists a unique $m \geq 0$ such that $\{P(S)\}^m \supseteq I$ but $\{P(S)\}^{m+1} \nsupseteq I$; by (c), $Sz_0^{m+1} \subset I \subseteq Sz_0^m$. Consider $M = Sz_0^m/Sz_0^{m+1}$ canonically as a left module over the pli-domain $D = S/P(S)$. Since ${}_DM$ is cyclic and D is a pli-domain, it follows that I/Sz_0^{m+1} is a cyclic left D-module. If $x_o \in I$ with $D(x_o + \{P(S)\}^{m+1}) = I/Sz_0^{m+1}$ then it is easily seen that $I = Sx_o$. Hence S is a completely primary pli-ring.

Suppose S is a completely primary pli-ring. Then (a) and (b) trivially hold. If S has large prime radical then (c) is contained in proposition 7.3. If S has small radical then we have to show that $P(S) \subseteq Ss$ for every $s \notin P(S)$; observe that for such an element s,

$s + P(S)$ is regular in $S/P(S)$; now Theorem 5.1 shows that $P(S) \subseteq Ss$. This completes the proof.

7.5. Proof of Theorem 7.1. Let R be a primary pli-ring with large prime radical such that $R/P(R) \cong M_n(D)$ where D is a pli-domain. By Proposition 7.3, $R \cong M_m(S)$ where S is a completely primary ring satisfying conditions (a) and (c) of 7.4. Clearly, $R/P(R) \cong M_m(S/P(S))$ where $S/P(S)$ is a domain. By Robson's Theorem (2.2), $m = n$ and $S/P(S) \cong D$, a pli-domain. Thus, 7.4 is applicable to S and shows that S is a completely primary pli-ring. This shows that (1) => (2). Trivially, (2) => (1). The uniqueness of n follows from 1.9. The uniqueness of S upto isomorphism follows from 2.8, 1.7 and 1.6. This completes the proof.

We now prove some results which will be used in the proof of Theorem 7.2.

7.6. LEMMA. Let R be a ring. Put $P = P(R)$. R is a completely primary pli-ring if and only if the following conditions are satisfied.

(1) P is nilpotent and $D = R/P$ is a pli-domain.

(2) Let $k+1$ be the index of nilpotency of P, where $k \geq 0$. Let B_n denote P^n/P^{n+1} $(1 \leq n \leq k)$ regarded canonically as a D-bimodule. Then, for $1 \leq n < k$, B_n is a divisible right D-module and a free left D-module of rank 1. B_k is a divisible right D-module and a non-zero cyclic left D-module.

(3) If I is a nilpotent left ideal of R such that $I \subseteq P^n$ but $I \not\subseteq P^{n+1}$ then $P^{n+1} \subseteq I$.

Proof. We shall firstly show that a ring R satisfying conditions (1), (2) and (3) is a completely primary pli-ring. We put an induction on k, where $k+1$ is the index of nilpotency of $P(R)$. If $k=0$, R is a pli-domain. Assume that the lemma is true for all rings satisfying conditions (a), (b), (c) and having the index of nilpotency of their prime radical $\leq k$; let R be as above. Then R/P^k satisfies conditions (a), (b), (c) and the index of nilpotency of $P(R/P^k) = P/P_k$ is k. By the induction hypothesis, R/P^k is a completely primary pli-ring.

Let I be a nilpotent left ideal of R. If $I \subseteq P^k$ then I is clearly a principal left ideal of R. If $I \nsubseteq P^k$, then using condition (3), we have $P^k \subseteq I$. Since R/P^k is a pli-ring, I/P^k is principally generated. If $I/P^k = (R/P^k)(x + P^k)$ then it is easily seen that $I = Rx$. Thus all nilpotent left ideals of R are principal. Let I be a non-nilpotent left ideal of R. Evidently, $I/P^k = (R/P^k)(c + P^k)$ for some $c \in R$, $c \notin P$. Since R/P is a domain, $c + P$ is a regular element of R/P. Thus, $P^k = P^k c \subseteq Pc$ since ${}_D P^k$ is a divisible module. By 7.4, $P/P^k \subseteq I/P^k$. It is now easy to see that $I = Rc$. This shows that R is a completely primary pli-ring and completes the induction on k.

Suppose that R is a completely primary pli-ring. Then (1) is evident and (3) is contained in 7.4. If $T(R) = (0)$ then by 6.7, 6.5, and 6.3, ${}_D B_n$ are free left D-modules of rank 1 for $1 \leq n \leq k$. If $T(R) \neq (0)$ then by 6.1, $T(R) = P^k$ and $T(R/T(R)) = (0)$. So, as above, ${}_D B_n$ are free left D-modules for $1 \leq n < k$; ${}_D B_k \cong {}_D P^k$ is clearly cyclic. The divisibility of B_{n_D} is an immediate consequence of 5.1. This completes the proof.

7.7. PROPOSITION. Let R be a ring. The following conditions on R are equivalent:

(1) R is a completely primary pli-ring.

(2) $G_{P(R)}(R)$ is a completely primary pli-ring.

Proof. (1) => (2). If R has large prime radical, by 6.2, $G_{P(R)}(R)$ is a primary pli-ring. Since $R/P(R) = G_{P(R)}(R)/P(G_{P(R)}(R))$, it follows that $G_{P(R)}(R)$ is a completely primary pli-ring. If R

has small prime radical, then using 6.12, it is easily seen that $G_{P(R)}(R)$ is a completely primary pli-ring with small prime radical.

(2) => (1). We shall show that R satisfies conditions (1), (2) and (3) of 7.6. Observe that

$$R/P(R) = G_{P(R)}(R) \; / \; P(G_{P(R)}(R)) = D \text{ say.}$$

So, D is a pli-domain. Since $P(G_{P(R)}(R))$ is nilpotent, it follows that $P(R)$ is nilpotent; further, they have the same index of nilpotency, say $k+1$. Also,

$$\{P(R)\}^n/\{P(R)\}^{n+1} = \{P(G_{P(R)}(R))\}^n/\{P(G_{P(R)}(R))\}^{n+1}$$

as D-bimodules for $1 \leq n \leq k$. Applying 7.6 to $G_{P(R)}(R)$, it follows that R satisfies conditions (1) and (2) of 7.6.

Since $G_{P(R)}(R)$ is a pli-ring, there exists $w_o \in P(R)$ such that

$$P(G_{P(R)}(R)) = G_{P(R)}(R) \; (w_o + \{P(R)\}^2).$$

Then $P(R) = R\,w_o + \{P(R)\}^2 = R\,w_o + \{R\,w_o + \{P(R)\}^2\}^2$ $= R\,w_o + \{P(R)\}^3$ etc. Since $P(R)$ is nilpotent, we have $P(R) = Rw_o$; so, $\{P(R)\}^n = R\,w_o^n$ for every $n \in Z^+$.

Let $c \in R$, $c \notin P(R)$. Since $\{P(R)\}^k_D$ is divisible, we have $\{P(R)\}^k = \{P(R)\}^k c \subseteq Rc$. Suppose $n+1 \leq k$ and $\{P(R)\}^{n+1} \subseteq Rc$. Divisibility of $\{P(R)\}^n/\{P(R)\}^{n+1}$ as a right D-module shows that

$$\{P(R)\}^n \subseteq \{P(R)\}^n c + \{P(R)\}^{n+1} \subset Rc.$$

It follows that $P(R) \subset Rc$.

Now, let I be a left ideal of R such that $I \subseteq \{P(R)\}^m$ but $I \nsubseteq \{P(R)\}^{m+1}$. Pick $w \in I \setminus \{P(R)\}^{m+1}$. Then $w = cw_o^n$ where $c \in R \setminus P(R)$. We now have

$$I \supseteq Rw = Rc\,w_o^n \supseteq Rw_o^{n+1} = \{P(R)\}^{n+1}.$$

Lemma 7.6 shows that R is a completely primary pli-ring.

7.8. Proof of Theorem 7.2. Proposition 7.3 shows that $R \cong M_n(S)$, where S is a completely primary ring. Clearly

$$G_{P(R)}(R) \cong M_n(G_{P(S)}(S)).$$

It is readily seen that $G_{P(S)}(S)$ is a completely primary ring. By Theorem 7.1, $G_{P(S)}(S)$ is a completely primary pli-ring. By 7.7, S is a completely primary pli-ring. This completes the proof.

References.

The results of this section are due to the author. Some of them may be found in Jategaonkar [1].

§8. PLI-RINGS WITH ADDITIONAL CONDITIONS.

A primary pli-ring with non-zero prime radical need not necessarily be a left Artinian ring; however, with some mild restrictions on the right ideals, they turn out to be so. More specifically, the following holds:

8.1. GOLDIE'S FOURTH THEOREM. Let R be a ring. The following conditions on R are equivalent:

(i) R is a pli-ring with the ascending chain condition on nilpotent right ideals.

(ii) R is a pli-ring with the descending chain on nilpotent right ideals.

(iii) R is a direct sum of a finite number of pli-rings, each of which is either a prime pli-ring or a primary left and right Artinian ring.

Proof. It is well-known that a left (resp. right) Artinian ring is left (resp. right) Noetherian. Lambek [1, page 69]. Thus, (iii) => (i) and (iii) => (ii). We shall simultaneously prove that (i) => (iii) and (ii) => (iii). Due to Johnson's first theorem, we may assume that R is a primary pli-ring with non-zero prime radical satisfying (i) or (ii) as the case may be. To get (iii), it then suffices to show that $R/P(R)$ is a simple Artinian ring. For, then each $\{P(R)\}^n/\{P(R)\}^{n+1}$ will be a left and right Artinian module over $R/P(R)$; so we can form a composition chain of left ideals of R and a composition chain of right ideals of R by lumping together the corresponding composition series of the modules $\{P(R)\}^n/\{P(R)\}^{n+1}$. Now, notice that $R/P(R) \cong \tilde{R}/P(\tilde{R})$, where $\tilde{R} = R/\{P(R)\}^2$. Further, $\tilde{R}$ satisfies (i) (resp. (ii)) if R satisfies (i) (resp. (ii)). To sum up, it suffices to prove the following claim: Let R be a primary pli-ring with non-zero prime radical satisfying the ascending or the descending chain condition on nilpotent right ideals. Let $\{P(R)\}^2 = (0)$. Then $R/P(R)$ is a simple Artinian ring. We proceed to prove this claim.

By 1.2 and 2.1, $R \cong M_n(S)$, where S is a completely primary left Noetherian ring with $\{P(S)\}^2 = (0)$. Let $D = S/P(S)$. Clearly, D is a left Noetherian domain, so has a l.q. skew field, say K. Since $\{P(S)\}^2 = (0)$, we may regard $P(S)$ canonically as a D-bimodule. Using 5.1, it is easily seen that $P(S)_D$ is a torsion-free divisible right D-module. Further, $P(S)_D$ has the ascending or descending chain condition on right D-submodules if R has the ascending or descending chain condition on nilpotent right ideals of R respectively. Since $P(S)_D$ is torsion-free and divisible, the domain of right operators of $P(S)_D$ can be extended from D to K so as to make $P(S)$ a right K-module. Let V_K be a one-dimensional right K-subspace of $P(S)_K$. The right D-submodules of V_K satisfy the ascending (resp. descending) chain condition if $P(S)_D$ has the ascending (resp. descending) chain condition on right D-submodules. However, V_K is isomorphic with the canonical right K-module K_K; it follows that K_D as a canonical right D-module has the ascending (resp. descending) chain condition on D-submodules.

Suppose K_D has the ascending chain condition on D-submodules. Let $\alpha \in D$, $\alpha \neq 0$. Consider the ascending chain

$$\alpha^{-1} D \subseteq \alpha^{-2} D \subseteq \cdots\cdots \subseteq \alpha^{-n} D \subseteq \cdots\cdots$$

of right D-submodules of K. It follows that $\alpha^{-(m+1)} D = \alpha^{-m} D$ for some $m \in Z^+$. Thus, $\alpha^{-(m+1)} = \alpha^{-m} d$ for some $d \in D$, so $\alpha^{-1} = d \in D$. Hence $K = D$.

Suppose K_D has the descending chain condition. Then D_D has the descending chain condition. Since D is a domain, it must be a skew field. So, $K = D$.

Since $R/P(R) \cong M_n(D)$, we have established our claim. This completes the proof.

8.2. PROPOSITION. A ring R is a direct sum of a finite number of pli-rings each of which is either a prime ring or a primary left Artinian ring if and only if R is a pli-ring with the descending chain condition on nilpotent left ideals and has a left Artinian l.q. ring.

Proof. In view of Johnson's first theorem, to prove the 'if' part it suffices to show that if R is a primary pli-ring with non-zero prime radical, has the descending chain condition on nilpotent left ideals and has a left Artinian l.q. ring then R is itself a left Artinian ring. However, 6.3, 6.4 and 6.5 show that ${}_{\bar{R}}\{P(R)\}^k \cong {}_{\bar{R}}\bar{R}$ where $k+1$ = index of nilpotency of $P(R)$ and $\bar{R} = R/P(R)$. It follows that $\bar{R}$ is left Artinian so R is left Artinian. The 'only if' part follows from Goldie's first theorem. This completes the proof.

8.3. COROLLARY. If R is a primary pli-ring with large prime radical satisfying the descending chain condition on nilpotent left ideals then R is left Artinian.

Proof. Use 6.1 and 8.2.

We remark that the corollary does not hold for primary pli-rings with small prime radical.

8.4. PROPOSITION. Let R be a primary pli-ring with large prime radical. If R has a right quotient ring then R is its own right quotient ring.

Proof. 1. Firstly, let R be a primary pli-ring with large prime radical having a left Artinian l.q. ring. (For the present, we do not assume that R has a r.q. ring). Let $P(R) = Rw_o$, $\overline{R} = R/P(R)$ and let Q be the simple Artinian l.q. ring of $\overline{R}$. Let $k+1$ = index of nilpotency of $P(R)$. Suppose c, c_1 are regular elements in R and $cr_1 = w_o^k c_1$ for some $r_1 \in R$. We claim that c is a unit in R.

By 6.3, 6.4 and 6.5, $Rw_o^k = \{P(R)\}^k$ is $\overline{R}$-isomorphic with the $(id., \rho|\overline{R})$ - $\overline{R}$ - bimodule, where $\rho: Q \to \overline{R}$ is a monomorphism. Thus, $w_o^k c_1 = c_2 w_o^k$ where $\rho(c_1 + P(R)) = c_2 + P(R)$. Clearly, $(c + Rw_o^k)(r_1 + Rw_o^k) = 0 + Rw_o^k$. By 6.11, $R/\{P(R)\}^k$ has a left Artinian l.q. ring; so by 5.1 and 5.3, $c + Rw_o^k$ is regular in $R/\{P(R)\}^k$. It follows that $r_1 \in Rw_o^k$, say $r_1 = r_2 w_o^k$, $r_2 \in R$. So,

$$(cr_2 - c_2)\, w_o^k = 0.$$

By 6.5, $cr_2 - c_2 \in P(R)$. Thus, in $\overline{R}$, we have

$$(c + P(R))(r_2 + P(R)) = c_2 + P(R) = \rho(c_1 + P(R)).$$

Clearly, $\rho(c_1 + P(R))$ is a unit in $\overline{R}$. So, $c + P(R)$ is a unit in $\overline{R}$. It follows that c is a unit in R. This proves our claim.

2. Now let R be as stated in the proposition. If $T(R) = (0)$ then R has a left Artinian l.q. ring and the above argument shows that every regular element of R is a unit in R. Thus R is its own r.q. ring. (In fact, R is left Artinian).

Suppose $T(R) \neq (0)$. Let c be a regular element in R. Then there exist $c_1, r_1 \in R$, c_1 regular in R, such that $cr_1 = w_o^{k-1} c_1$. Theorems 6.1, 5.1 and 5.3 show that the first part of the proof is applicable to the equation

$$(c + T(R))(r_1 + T(R)) = (w_o^{k-1} + T(R))(c_1 + T(R))$$

in $R/T(R)$. Thus $c + T(R)$ is a unit in $R/T(R)$. Consequently, c is a unit in R. This completes the proof.

8.5. COROLLARY. Let R be a primary pli-ring with non-zero prime radical. If a left Artinian ring Q is a l.q. ring and right quotient ring of R then Q = R.

Proof: clear.

References.

Theorem 8.1 is a generalization of a theorem of Goldie [3]. Our proof is new. The rest is from Jategaonkar [1].

CHAPTER III

IDEAL THEORY OF FULLY LEFT GOLDIE IPLI-RINGS

INTRODUCTION

In this chapter, we develope the ideal theory of pli-rings (more generally, of fully left Goldie ipli-rings). Our theory is a caricature of the Dedekind ideal theory.

§§1 and 2 show that all features of the Dedekind ideal theory go over except that there may be embedded primes. Examples of this sort are available in Jategaonkar [3,4]. In §3, we use the results of §§1, 2 to describe in detail the make up of local pli-rings. §4 shows that the existence of embedded primes is a one-sided affair; under symmetric hypothesis, the classical ideal theory holds.

§1. IDEAL THEORY - I.

This section is devoted to a study of ideals in a fully left Goldie ipli-ring. Due to Johnson's first theorem, it clearly suffices to restrict our attention to ideals in a primary fully left Goldie ipli-ring.

We need some definitions. An ideal I of a ring R is called a primary ideal of R if, for two-sided ideals A, B of R, $A \cdot B \subseteq I$ implies either $A \subseteq I$ or $B^n \subseteq I$ for some $n \in Z^+$. (In the literature, primary ideals as defined above are sometimes called right primary ideals; left primary ideals are then defined in an analoguous manner. However, since we shall not need the concept of left primary ideals, no confusion is likely by our use of the term 'primary ideals' as defined above.) Clearly, I is a primary ideal of R if and only if R/I is a primary ring.

Let I be a proper ideal of a ring R. The prime radical of I in R, denoted by $P_R(I)$ or $P(I)$, is the intersection of all those prime ideals of R which contain I. Thus, by definition, $P(R) = P_R(0)$.

We now define transfinite powers of an ideal A of a ring R. One definition is the following:

$$A^1 = A;$$

$$A^\alpha = A \cdot A^\beta \quad \text{if} \quad \alpha = \beta+1;$$

$$A^\alpha = \bigcap_{\beta<\alpha} A^\beta \quad \text{if} \quad \alpha \text{ is a limit ordinal.}$$

We shall need this definition later on. However, for the present, we shall need another definition of transfinite powers of ideals of R; this definition is notationally distinguished from the first by writing the index ordinal in a square bracket; it runs as follows:

$$A^{[\omega^0]} = A;$$

$$A^{[\omega^\alpha]} = \bigcap_{n\in Z^+} \{A^{[\omega^\beta]}\}^n \quad \text{if} \quad \alpha = \beta+1;$$

$$A^{[\omega^\alpha]} = \bigcap_{\beta<\alpha} \{A^{[\omega^\beta]}\} \quad \text{if } \alpha \text{ is a limit ordinal.}$$

Notice that the second definition is defined only for index ordinals of the form ω^α, where ω is the first infinite ordinal and α is an arbitrary ordinal.

We now state the main results of this section.

1.1. THEOREM. Let R be a fully left Goldie ipli-ring and A be a proper prime ideal of R. Then there exists an ordinal α such that $P(R) = A^{[\omega^\alpha]}$. If $P(R) \neq (0)$ then α is uniquely determined by A.

Let α be the first ordinal such that $P(R) = A^{[\omega^\alpha]}$. Then $A^{[\omega^\beta]} \subsetneq A^{[\omega^\gamma]}$ for $\gamma < \beta \leq \alpha$. The prime ideals of R contained in A are precisely those of the form $A^{[\omega^\beta]}$ for $\beta \leq \alpha$.

1.2. COROLLARY. Let A be a proper non-zero ideal in a primary fully left Goldie ipli-ring R.

(a) A is a prime ideal if and only if $A = M^{[\omega^\alpha]}$ for some maximal ideal M and some ordinal α.

(b) If A is a prime ideal of R and M is a maximal ideal of R containing A, then there exists a unique ordinal α such that $A = M^{[\omega^\alpha]}$.

1.3. THEOREM. Let A be a proper ideal in a primary fully left Goldie ipli-ring. Then the following conditions are equivalent:

(i) A is a primary ideal of R.

(ii) $P(A)$ is a prime ideal of R.

(iii) There exists an integer $n \in Z^+$ and a prime ideal B of R such that $B^n \subseteq A \subseteq B$. (In fact, $B = P(A)$).

1.4. THEOREM. Let A be a proper ideal in a primary fully left Goldie ipli-ring R. Then there exist primary ideals A_i, $1 \leq i \leq n$, of R such that

$$P(A_i) \nsubseteq P(A_j) \quad \text{if } i \neq j;$$

$$A_i A_j = A_j A_i \quad \text{for all } i, j;$$

and

$$A = A_1 : \ldots : A_n.$$

Further, these ideals are uniquely determined by A upto order.

1.5. THEOREM. Let A, B be proper primary ideals of a primary fully left Goldie ipli-ring R.

(i) If $P(A) \nsubseteq P(B)$ and $P(B) \nsubseteq P(A)$ then $AB = BA = A \cap B$; further, AB is not a primary ideal of R.

(ii) If $P(A) = P(B)$ then AB and BA are primary ideals of R and $P(AB) = P(BA) = P(A)$.

(iii) If $P(A) \subsetneq P(B)$ then $AB = A$. Also, BA is a primary ideal with $P(BA) = P(A)$.

Notice that theorem 1.4 expresses every ideal in a primary fully left Goldie ipli-ring as a product of a finite number of commuting primary ideals in an essentially unique way, theorem 1.3 characterizes primary ideals in terms of prime ideals and theorem 1.1 and corollary 1.2 express prime ideals in terms of maximal ideals. Theorem 1.5 contains some information as to how primary ideals multiply.

The main aspect in which the ideal theory described above differs from that of commutative PIR's (or left-right principal ideal rings) is that it allows for the existence of non-zero non-maximal primes. Illustrative examples of this phenomenon can be found in Jategaonkar [3,4].

We now prove the theorems (in the order in which they are stated). Our main tools are the three theorems of R. E. Johnson proved in §4 of chapter II.

1.6. LEMMA. Let $\{A_\gamma: \gamma \in \Gamma\}$ be a non-empty set of prime ideals of a ring R. If $\{A_\gamma: \gamma \in \Gamma\}$ is totally ordered under inclusion then $A = \bigcap_{\gamma\in\Gamma} A_\gamma$ is a prime ideal of R.

Proof. Clearly, A is an ideal of R. Suppose $a, b \in R$, $b \notin A$ and $aRb \subseteq A$. There exists $\gamma_0 \in \Gamma$ such that $b \notin A_{\gamma_0}$. Thus, for every $A_\gamma \subseteq A_{\gamma_0}$, we have $aRb \subseteq A_\gamma$ but $b \notin A_\gamma$; since A_γ is a prime ideal of R, it follows that $a \in A_\gamma$. Since $A = \bigcap \{A_\gamma | A_\gamma \subseteq A_{\gamma_0}\}$, we have $a \in A$. This proves the lemma.

For the rest of this section, R denotes a primary fully left Goldie ipli-ring.

1.7. LEMMA. If A is a non-zero proper ideal of R then $A^2 \subset A$.

Proof. Suppose $A^2 = A = Ra \neq (0)$. Then A is a non-nilpotent ideal of R. By Johnson's third theorem $\ell(a) = (0)$. Since $Ra^2 = A^2 = Ra$, there exists $r \in R$ such that $(1-ra)a = 0$; so, $1 = ra \in Ra$ so $A = R$, a contradiction. Hence $A^2 \subset A$. This proves the lemma.

1.8. LEMMA. If A is a prime ideal of R then A^n is a primary ideal of R for every $n \in Z^+$.

Proof. Let $n \in Z^+$. Put $S = R/A^n$ and $\bar{A} = A/A^n$. Then $S/\bar{A} = (R/A^n)/(A/A^n) \cong R/A$, a prime ring. Clearly, $\bar{A}^n = (\bar{0})$. Thus, $P(S) = \bar{A}$. Johnson's second theorem shows that S is a primary ring; so, A^n is a primary ideal of R. This completes the proof.

We now prove the following crucial lemma.

1.9. LEMMA. Let A be a non-nilpotent proper prime ideal of R. Let $B = \bigcap_{n\in Z^+} A^n$. Then $B \subset A$ and B is a prime ideal of R. Further, if A' is a prime ideal of R with $A' \subset A$ then $A' \subseteq B$.

<u>Proof</u>. Using lemma 1.7, it is easily seen that $\{A^n: n \in Z^+\}$ is a strictly decreasing sequence of ideals of R. Thus, $B \subset A$.

Let C, D be two-sided ideals of R such that $CD \subseteq B$ but $D \nsubseteq B$. Let us select D maximally: $D = \{r \in R | C r \subseteq B\}$. Let $A = Ra$ so that, by Johnson's theorem, $\ell(a) = (0)$. Let $x \in R$ with $x a \in D$. Then, for every $n \geq 2$, $C x a \subseteq A^n = Ra^n$; so, for every $c \in C$, there exists $r \in R$ such that $(cx - ra^{n-1}) a = 0$; since $\ell(a) = (0)$, we have $cx = ra^{n-1} \in A^{n-1}$. Thus $Cx \in A^{n-1}$ for every $n \geq 2$. Consequently, $Cx \subseteq B$ so that $x \in D$. Hence $D = \{x \in R | x a \in D\}$.

If possible, let $D \subseteq A$. Since $D \nsubseteq B$ and since $\{A^n: n \in Z^+\}$ is a strictly decreasing sequence of ideals of R, we may choose $k \in Z^+$ maximally such that $D \subseteq A^k$. Let $D = Rd$. Then $d = \bar{d} a^k$; so, as shown above, $\bar{d} \in D$. However, then $\bar{d} \in A^k$ and $Rd \subseteq A^{2k}$, contrary to our choice of k. Hence $D \nsubseteq A$. Since A is a prime ideal, we have $D^m \nsubseteq A^n$ for all $m, n \in Z^+$. Now, as shown in lemma 1.8, A^n is a primary ideal of R, $CD \subseteq A^n$ and $D^m \nsubseteq A^n$. for all $m \in Z^+$; so, $C \subseteq A^n$ for every $n \in Z^+$. Consequently, $C \subseteq B$. We have thus shown that B is a prime ideal.

Let $A' = Ry$ be a prime ideal of R such that $Ry \subset Ra$. If possible, let $Ry \nsubseteq B = \bigcap_{n \in Z^+} Ra^n$. Let t be the largest integer such that $Ry \subseteq Ra^t$; then there exists $r \in R$ such that $r \notin Ra$ and $y = ra^t$. Let $\bar{R} = R/Ry$ and let $\psi: R \to \bar{R}$ be the canonical epimorphism. Since Ry is a proper prime ideal of R, $\bar{R}$ is a prime fully left Goldie ipli-ring. Since $Ry \subset Ra$, therefore $\bar{R}\,\psi(a)$ is a non-zero ideal of $\bar{R}$. By Johnson's third theorem, $\psi(a)$ is left regular in $\bar{R}$. Now, $y = ra^t$ implies $\psi(0) = \psi(r)\,[\psi(a)]^t$ so that $\psi(0) = \psi(r)$ and $r \in Ry \subseteq Ra^t$, contrary to $r \notin Ra$. Hence $A' \subseteq B$. This completes the proof.

1.10. Proof of theorem 1.1. Let A be a proper prime ideal of R. If possible, let $P(R) \neq A^{[\omega^\alpha]}$ for every ordinal α. It follows from lemmas 1.6 and 1.9 that $A^{[\omega^\alpha]}$ is a prime ideal of R; so, $P(R) \subset A^{[\omega^\alpha]}$ for every ordinal α. Let η be an ordinal of cardinality greater than the cardinality of R. Then, by lemma 1.9, $\{A^{[\omega^\alpha]}: \alpha \leq \eta\}$ is a strictly decreasing chain of ideals of R, contrary to our choice of η. Hence $P(R) = A^{[\omega^\alpha]}$ for some ordinal α.

Suppose $P(R) \neq (0)$ and $P(R) = A^{[\omega^\alpha]} = A^{[\omega^\beta]}$ with $\beta < \alpha$. Then nilpotency of $P(R)$ implies $A^{[\omega^{\beta+1}]} = (0)$. However, it is clear from our definition of transfinite powers that $A^{[\omega^\alpha]} \subseteq A^{[\omega^{\beta+1}]}$. Thus, $A^{[\omega^\alpha]} = (0) = P(R)$, contrary to our assumption.

Now, let α be the first ordinal such that $P(R) = A^{[\omega^\alpha]}$. (We no longer assume that $P(R) \neq (0)$). A transfinite induction using lemmas 1.6 and 1.9 shows that $A^{[\omega^\beta]}$ is a prime ideal of R for $\beta \leq \alpha$ and $A^{[\omega^\beta]} \subset A^{[\omega^\gamma]}$ if $\gamma < \beta \leq \alpha$.

Let A' be a non-zero prime ideal of R contained in A. If $A' \subseteq A^{[\omega^\mu]}$ then clearly $A' \subseteq A^{[\omega^\lambda]}$ for every $\lambda \leq \mu$. Thus the set $\Omega = \{\lambda \leq \alpha\colon\ A' \subseteq A^{[\omega^\lambda]}\}$ is an initial segment of the set of all ordinals $\leq \alpha$. Further if β is a limit ordinal, $\beta \leq \alpha$ and if $\gamma \in \Omega$ for every $\gamma < \beta$ then $A' \subseteq \bigcap_{\gamma<\beta} A^{[\omega^\gamma]} = A^{[\omega^\beta]}$; so $\beta \in \Omega$. It follows that Ω has a last element, say ξ. Clearly $\xi \leq \alpha$ and $A' \subseteq A^{[\omega^\xi]}$. If $A' \subset A^{[\omega^\xi]}$ then, by 1.9, $A' \subseteq A^{[\omega^{\xi+1}]}$, contrary to our choice of ξ. Hence $A' = A^{[\omega^\xi]}$. This proves the theorem.

1.11. Proof of corollary 1.2. Follows easily from theorem 1.1.

1.12. Proof of theorem 1.3. We give a cyclic proof.

(i) => (ii). Suppose A is a primary ideal of R. Then R/A is a primary fully left Goldie ipli-ring; so, by Johnson's second theorem $P(R/A)$ is a prime ideal of R/A and every prime ideal of R/A contains $P(R/A)$. Thus if $\sigma\colon\ R \to R/A$ denotes the canonical epimorphism, then $\sigma^{-1}(P(R/A))$ is a prime ideal of R which contains A and is contained in every prime ideal of R containing A. i.e., $P(A) = \sigma^{-1}(P(R/A))$, a prime ideal of R.

(ii) => (iii). Clearly, $P(R/A) = P(A)/A$. Since $P(R/A)$ is nilpotent, it follows that $\{P(A)\}^n \subseteq A \subseteq P(A)$ for some $n \in Z^+$.

(iii) => (i). In R/A, B/A is a nilpotent prime ideal of R. By Johnson's second theorem, R/A is a primary ring; so, A is a primary ideal of R. This completes the proof.

<u>1.13. LEMMA</u>. If A, B are proper ideals of R such that $A + B = R$ then $A \cap B = AB = BA$.

<u>Proof</u>. Let $A=Ra$, $B=Rb$, $\bar{R}=R/AB$, $\bar{A}=A/AB$, $\bar{B}=B/AB$, $\bar{a}=a+AB$ and $\bar{b} = b + AB$. Then $\bar{R} = \bar{A} + \bar{B}$ and $\overline{AB} = (\bar{0})$ so that $\overline{ab} = \bar{0}$. Since $\bar{B} = (\bar{A} + \bar{B})\bar{B} = \bar{B}^2$. By lemma 1.2.5, we have $\ell_{\bar{R}}(\bar{b}) \cap \bar{B} = (\bar{0})$. So, $\bar{B}\cdot\ell_{\bar{R}}(\bar{b}) = (\bar{0})$. i.e., $\ell_{\bar{R}}(\bar{b}) \subseteq r_{\bar{R}}(\bar{b})$. Now, $\overline{ab} = \bar{0}$ gives $\bar{b}\,\bar{a} = \bar{0}$ so that $\bar{B}\,\bar{A} = (\bar{0})$ and $BA \subseteq AB$. Similarly, $AB \subseteq BA$. Consequently, $AB = BA$. Now, $AB + BA \subseteq A \cap B = (A+B)(A\cap B) = A(A\cap B) + B(A\cap B) \subseteq AB + BA$ shows that $A\cap B = AB + BA = AB = BA$. This completes the proof.

<u>1.14. LEMMA</u>. Let $n \in Z^+$ and let Ra_i, $1 \leq i \leq n+1$, be proper primary ideals of R such that $P(Ra_i) \nsubseteq P(Ra_j)$ for $i \neq j$. Then

$$(Ra_1 \vdots \ldots \vdots Ra_n) \cap Ra_{n+1} = Ra_1 \vdots \ldots \vdots Ra_n\, Ra_{n+1}.$$

Also, $Ra_1 \vdots \ldots \vdots Ra_n\, Ra_{n+1}$ is not a primary ideal of R.

<u>Proof</u>. Put $a = a_1 \vdots \ldots \vdots a_n$ so that $Ra = Ra_1 \vdots \ldots \vdots Ra_n$. We claim that $Ra + Ra_{n+1} = R$. If not, let M be a maximal ideal of R containing both Ra and Ra_{n+1}. Since M is a prime ideal and $Ra_1 \vdots \ldots \vdots Ra_n \subseteq M$; therefore $Ra_i \subseteq M$ for at least one i, $1 \leq i \leq n$. Since Ra_i and Ra_{n+1} are both primary ideals, using theorem 1.3, it follows that $P(Ra_i)$ and $P(Ra_{n+1})$ are prime ideals contained in M. By corollary 1.2, we obtain an inclusion relation among $P(Ra_i)$

and $P(Ra_{n+1})$, contrary to our hypothesis. This proves one claim. Lemma 1.13 now shows that

$$(Ra_1 : \ldots : Ra_n) \cap Ra_{n+1} = Ra_1 : \ldots : Ra_n \; Ra_{n+1}.$$

Let $S = R/Raa_{n+1}$ and let $\sigma\colon R \to S$ be the canonical epimorphism. S is clearly a fully left Goldie ipli-ring. Suppose S is primary. Then $S\sigma(a)S\sigma(a_{n+1}) = (\sigma(0))$ implies that either $S\sigma(a) = (\sigma(0))$ or $S\sigma(a^m_{n+1}) = (\sigma(0))$ for some $m \in Z^+$. If $S\sigma(a) = (\sigma(0))$ then $Ra \subseteq Ra\; Ra_{n+1} \subseteq P(Ra_{n+1})$. By 1.3, $P(Ra_{n+1})$ is a prime ideal. It follows that $P(Ra_i) \subseteq P(Ra_{n+1})$ for some i, $1 \leq i \leq n$; this is contrary to our hypothesis. If $S\sigma(a^m_{n+1}) = \sigma(0)$ then $Ra^m_{n+1} \subseteq Ra\; Ra_{n+1} \subseteq P(Ra_1)$; so, $P(Ra_{n+1}) \subseteq P(Ra_1)$, a contradiction again. Hence S is not a primary ring i.e., $Ra\; Ra_{n+1}$ is not a primary ideal of R. This completes the proof.

<u>1.15. Proof of theorem 1.4</u>. Let A be a proper ideal of R and $S = R/A$. Clearly, S is a fully left Goldie ipli-ring, so that, by Johnson's first theorem,

$$S/A = S_1 \oplus \ldots\ldots \oplus S_n,$$

where $n \in Z^+$ and each S_i, $1 \leq i \leq n$, is a primary fully left Goldie ipli-ring. Evidently,

$$\bar{A}_i = S_1 \oplus \ldots\ldots \oplus S_{i-1} \oplus S_{i+1} \oplus \ldots\ldots \oplus S_n$$

is a primary ideal of S for each $i = 1, \ldots\ldots, n$. Let $\sigma\colon R \to S$ be the canonical epimorphism and let $Ra_i = \sigma^{-1}(\bar{A}_i)$ for $1 \leq i \leq n$. Since $\bigcap_{i=1}^{n} \bar{A}_i = (\bar{0})$ and since $Ra_i \supseteq A$, it is clear that each Ra_i is a primary ideal and $A = \bigcap_{i=1}^{n} Ra_i$. Since

$$P(\bar{A}_i) = S_1 \oplus \ldots \oplus S_{i-1} \oplus P(S_i) \oplus S_{i+1} \oplus \ldots\ldots \oplus S_n,$$

and since $P(Ra_i) = \sigma^{-1}(P(\bar{A}_i))$, it follows that $P(Ra_i) \not\subseteq P(Ra_j)$ if

$i \neq j$. Lemma 1.14 now shows that $A = Ra_1 \dot{:} \ldots \dot{:} Ra_n$ and that Ra_i commute with each other under multiplication.

It remains to establish the assertion concerning uniqueness. Suppose the ideal A can also be expressed as

$$A = Rb_1 \dot{:} \ldots \dot{:} Rb_m,$$

where $m \in Z^+$, Rb_i are primary ideals of R which commute under multiplication for $1 \leq i \leq m$ and $P(Rb_i) \nsubseteq P(Rb_j)$ for $i \neq j$. Then, for any i, $1 \leq i \leq m$, we have

$$Ra_1 \dot{:} \ldots \dot{:} Ra_n \subseteq Rb_i \subseteq P(Rb_i).$$

Using theorem 1.3, we obtain an integer $k(i)$, $1 \leq k(i) \leq n$, such that $P(Ra_{k(i)}) \subseteq P(Rb_i)$. Similarly, for every $i = 1, \ldots, n$, there exists $\ell(i)$, $1 \leq \ell(i) \leq m$ such that $P(Rb_{\ell(i)}) \subseteq P(Ra_i)$. Hence $m = n$ and, after rearranging and reindexing if necessary, $P(Ra_i) = P(Rb_i)$ for $1 \leq i \leq n$.

Let $\overline{R} = R/Ra_1$ and let ψ: $R \to \overline{R}$ be the canonical epimorphism. Then

$$\psi(0) = \psi(A) = \psi(Rb_1 \dot{:} \ldots \dot{:} Rb_n).$$

Since $\overline{R}$ is a primary ring, either

$$\psi(Rb_1 \dot{:} \ldots \dot{:} Rb_{n-1}) = \psi(0) \quad \text{or} \quad \psi(Rb_n^t) = \psi(0)$$

for some $t \in Z^+$. In the later case, we have $Rb_n^t \subseteq Ra_1 \subseteq P(Ra_1) = P(Rb_1)$ so that, using 1.3, $P(Rb_n) \subseteq P(Rb_1)$, a contradiction. Thus

$$\psi(Rb_1 \dot{:} \ldots \dot{:} Rb_{n-1}) = (\psi(0)).$$

Repetition of this argument yields $\psi(Rb_1) = (\psi(0))$; so, $Rb_1 \subseteq Ra_1$. Similarly, $Ra_1 \subseteq Rb_1$. Hence $Ra_1 = Rb_1$. Since Ra_i commute with each other and Rb_i commute with each other, the above argument yields $Ra_i = Rb_i$ for $i = 1, \ldots, n$. This proves the theorem.

<u>1.16. Proof of theorem 1.5.</u> (i) is contained in lemma 1.14. (ii) follows easily from theorem 1.3.

Let A, B be primary ideals with $P(A) \subset P(B)$. By 1.3, $\{P(B)\}^m \subset B$ for some $m \in Z^+$. By 1.9, $P(A) \subseteq \bigcap_{n \in Z^+} \{P(B)\}^n \subseteq B$. By 1.3, $\{P(A)\}^k \subseteq A$ for some $k \in Z^+$. Thus,

$$\{P(A)\}^{k+1} \subseteq BA \subseteq P(A).$$

Theorem 1.3 now shows that BA is a primary ideal with $P(BA) = P(A)$. A similar argument shows that AB is a primary ideal with $P(AB) = P(A)$.

Let $A = Ra$ and $B = Rb$. Since $AB = Rab$ is a primary ideal of R, $S = R/Rab$ is a primary fully left Goldie ipli-ring. Let σ: $R \to S$ be the canonical epimorphism. Put $P(Ra) = Rc$ and $P(Rb) = Rd$. Then $S\sigma(c) = P(S)$. By 1.3 and 1.9, we have $Rc \subset Rd^m \subseteq Rb$ for some $m \in Z^+$. Consequently, $P(S) \subset S\sigma(b)$. By Johnson's third theorem, $\sigma(b)$ is a left regular element in S. Now $\sigma(ab) = \sigma(a)\sigma(b) = \sigma(0)$ implies $\sigma(a) = \sigma(0)$ i.e. $a \in Rab$. Hence $A = AB$. This completes the proof.

§2. PRIME FACTORIZATION.

We now consider ideal theory of a prime fully left Goldie ring in which every primary ideal is a product of a finite number of prime ideals. It is shown that the theory so obtained is applicable to prime fully left Goldie ipli-rings in which every maximal left ideal is two-sided; in particular, to local prime fully left Goldie ipli-rings.

We need some definitions. Let R be a primary fully left Goldie ipli-ring. Inductively, we define "admissible n-tuples of prime ideals" of R as follows: Any 1-tuple of prime ideals of R is admissible. Suppose admissible n-tuples of prime ideals of R have been defined for every $n < m$. An m-tuple $(P_1,\dots,P_m)$ of prime ideals of R is admissible if there exist $1 \leq i_1 < \dots < i_k = m$ such that

(a) $P_{i_s} \nsubseteq P_{i_t}$ for $s \neq t$;

(b) $P_{i_s} \subseteq P_j$ for $i_{s-1} + 1 \leq j \leq i_s$ where $i_o = 0$.

(c) If $i_{s-1} < i_s - 1$ then

$$(P_{i_{s-1}+1},\dots, P_{i_s-1})$$

is an admissible tuple.

The subtuples $(P_{i_{s-1}+1},\dots, P_{i_s})$, $1 \leq s \leq k$, are called the primary components of the admissible m-tuple $(P_1,\dots,P_m)$ of prime ideals of R. Notice that if $(P_1,\dots,P_m)$ is an admissible m-tuple then, by 1.5, the prime ideals which are entries of distinct primary components of $(P_1,\dots,P_m)$ commute with each other under multiplication. Also, if $(P_1,\dots,P_m)$ is admissible then so also is $(P_1,\dots,P_n)$ for $n \leq m$.

We now define "permissible rearrangements" of entries of an admissible n-tuple of prime ideals of R. For a 1-tuple, the identity rearrangement is permissible. Suppose permissible rearrangements of entries are defined for all admissible n-tuples with $n < m$. Let

$(P_1,\ldots,P_m)$ be an admissible m-tuple of prime ideals of R. If it has only one primary component viz. $(P_1,\ldots,P_m)$ then a rearrangement is permissible if it leaves P_m fixed and induces a permissible rearrangement on the admissible (m-1)-tuple $(P_1,\ldots,P_{m-1})$. If $(P_1,\ldots,P_m)$ has more than one primary components than a rearrangement is permissible if it can be expressed as a permutation of primary components followed by a rearrangement which induces permissible rearrangements on primary components. Note that if $(P_1,\ldots,P_m)$ is admissible then any m-tuple obtained from it by a permissible rearrangement is also admissible.

A non-zero element $p \in R$ is called a (strong) prime element if Rp is a proper prime ideal of R. A n-tuple $(p_1,\ldots,p_n)$ of prime elements is admissible if the n-tuple $(Rp_1,\ldots,Rp_n)$ of prime ideals of R is admissible; a rearrangement of entries of $(p_1,\ldots,p_n)$ is permissible if it induces a permissible rearrangement of entries of $(Rp_1,\ldots,Rp_n)$.

Let a,b be elements of a ring R. a and b are left associates in R if a = ub for some unit u in R. If R is a prime fully left Goldie ipli-ring and if Ra and Rb are non-zero ideals then, by Johnson's third theorem, a and b are left associates if and only if Ra = Rb.

2.1. THEOREM. Let R be a prime fully left Goldie ipli-ring in which every ideal is a product of a finite number of prime ideals of R. Let A be a non-zero proper ideal of R. Then there exists an admissible n-tuple $(P_1,\ldots,P_n)$ of non-zero prime ideals of R such that

$$A = P_1:\ldots:P_n .$$

Further, if $(Q_1,\ldots,Q_m)$ is an admissible m-tuple of prime ideals of R such that

$$A = Q_1:\ldots:Q_m$$

then n = m and $(Q_1,\ldots,Q_n)$ can be obtained from $(P_1,\ldots,P_n)$ by a permissible rearrangement of entries of $(P_1,\ldots,P_n)$.

2.2. THEOREM. Let R be a prime fully left Goldie ipli-ring in which every maximal left ideal is two-sided. Then R is a pli-domain, every left ideal of R is two-sided and every ideal of R is a product of a finite number of prime ideals of R.

Let a be a non-zero non-unit in R. Then there exists an admissible n-tuple $(p_1,\dots,p_n)$ of prime elements of R such that

$$a = p_1 \colon \dots \colon p_n \quad .$$

Further, $(p_1,\dots,p_n)$ is uniquely determined by a upto permissible rearrangements and associates.

2.3. Proof of theorem 2.1. Let A be a non-zero proper ideal of R. By our hypothesis, there exist prime ideals $P_1,\dots,P_n$ of R such that

$$A = P_1 \colon \dots \colon P_n \quad .$$

Let $P_{i_1},\dots,P_{i_k}$ be all the distinct prime ideals of R which are minimal in the set $\{P_j \colon 1 \leq j \leq n\}$. From theorem 1.1, it is clear that every P_j $(1 \leq j \leq n)$ contains precisely one of the ideals $P_{i_1},\dots,P_{i_k}$. If P_i and P_j contain distinct prime ideals out of $P_{i_1},\dots,P_{i_k}$ then, by theorem 1.1, there is no inclusion relation between P_i and P_j; so, by 1.5, $P_iP_j = P_jP_i$. Further, if $P_j \supsetneqq P_{i_t}$ then, by 1.5, $P_{i_t}P_j = P_{i_t}$. Consequently after rearranging and deleting some prime ideals from the given prime factorization of A and reindexing if necessary, we may assume that

$$A = P_1 \colon \dots \colon P_n$$

where $P_j \supseteq P_{i_s}$ for $i_{s-1} + 1 \leq j \leq i_s$, $1 \leq s \leq k$. An induction is now available to prove the existence part of the theorem.

Let $(P_1,\dots,P_k)$ and $(Q_1,\dots,Q_\ell)$ be admissible tuples of non-zero prime ideals of R such that

$$A = P_1 \colon \dots \colon P_k = Q_1 \colon \dots \colon Q_\ell \quad .$$

If $P_{i_1},\ldots,P_{i_s}$ (resp. $Q_{j_1},\ldots,Q_{j_t}$) are all the distinct prime ideals which are minimal in the set $\{P_i: 1 \leq i \leq k\}$ (resp. $\{Q_j: 1 \leq j \leq \ell\}$) then each P_{i_m} contains some Q_{j_n} and each Q_{j_n} contains some P_{i_m}. Since there are no inclusion relations among $P_{i_1},\ldots,P_{i_s}$ and no inclusion relations among $Q_{j_1},\ldots,Q_{j_t}$, it follows that $s = t$ and that there exists a bijection

$$\varphi:\{i_1,\ldots,i_s\} \to \{j_1,\ldots,j_t\}$$

such that $P_{i_m} = Q_{\varphi(i_m)}$ for $1 \leq m \leq s$. After a permissible rearrangement of entries of $(P_1,\ldots,P_k)$, we may assume that $\varphi(i_s) = j_s$. Noting that $i_s = k$ and $j_s = \ell$, we have $P_k = Q_\ell$.

Now, let $Rb = P_1:\ldots:P_{k-1}$, $Rc = Q_1:\ldots:Q_{\ell-1}$ and $P_k = Q_\ell = Rd$. We then have

$$A = Rbd = Rcd.$$

Thus $bd = ucd$ and $cd = vbd$ for some $u, v \in R$. So, $(1 - uv)bd = (1-vu)cd = 0$. By Johnson's third theorem, b,c and d are left regular in R. So, u is a unit in R. Now $(b-uc)d = 0$ gives $b = uc$ so $Rb = Rc$. Hence $P_1:\ldots:P_{k-1} = Q_1:\ldots:Q_{\ell-1}$. Notice that $(P_1,\ldots,P_{k-1})$ and $(Q_1,\ldots,Q_{\ell-1})$ are both admissible tuples of prime ideals of R. An induction is now available to conclude the proof.

We need a remark concerning partially ordered sets before we can prove theorem 2.2.

Let X be a non-empty set and $<$ be an anti-reflexive, transitive relation on X. Further, let $(X, <)$ satisfy the ascending chain condition viz. there does not exist a sequence $\{x_n: n \in Z^+\}$ such that $x_n < x_{n+1}$ for every $n \in Z^+$. Notice that every non-empty subset Y of X with the induced anti-reflexive transitive relation also satisfies the ascending chain condition; so, Y has at least one element which is maximal in $(Y, <)$.

We obtain a decomposition of X as follows: Let X_0 be the set of all maximal elements of X. If α is a non-zero ordinal and if X_β is defined for every $\beta < \alpha$ then X_α is the set of all maximal elements of $X \setminus (\bigcup_{\beta<\alpha} X_\beta)$. Inductively, this defines X_α for every ordinal α. An easy cardinality argument shows that there exists an ordinal τ for which $X_\tau = \phi$. If τ is the first ordinal with $X_\tau = \phi$ then clearly, $\{X_\beta : \beta < \tau\}$ is a decomposition of X into non-empty disjoint subsets. $\{X_\beta : B < \tau\}$ is called the stratification of X. For $\beta < \tau$, the set X_β is called the βth stratum of X.

2.4. Proof of theorem 2.2. We shall show that every non-zero left ideal Ra of R is a product of prime ideals.

Let X be the set of all proper prime ideals of R. Then $\subset$ is clearly a anti-reflexive, transitive relation on X and $(X,\subset)$ satisfies the ascending chain condition. Let $\{X_\beta : \beta < \tau\}$ be the stratification of X. We shall make a transfinite induction; our β^{th} transfinite induction hypothesis is the following: If $a \in R$ and $a \notin \cup\{P | P \in X_\beta\}$ then Ra is a product of a finite number of prime ideals of R.

Since every maximal left ideal of R is assumed to be two-sided, the 0th hypothesis is trivially true.

Let $0 < \alpha < \tau$ and let β^{th} hypothesis be true for every $\beta < \alpha$. Let $a \in R$, $a \notin \cup\{P | P \in X_\alpha\}$. We now have to show that Ra is a product of a finite number of prime ideals of R. If $a \notin \cup\{P | P \in X_\beta\}$ for some $\beta < \alpha$ then there remains nothing to prove. Suppose $a \in \cup\{P | P \in X_\beta\}$ for every $\beta < \alpha$. Let $RaR = Rb$. It is clear that $b \in \cup\{P | P \in X_\beta\}$ for every $\beta < \alpha$. Consider $P(Rb)$, the prime radical of Rb. Using Johnson's first theorem, it follows that

$$P(Rb) = P_1 \cap \ldots \cap P_n$$

where P_i are prime ideals of R and $P_i \nsubseteq P_j$ for $i \neq j$. A repeated application of lemma 1.14 yields

$$P(Rb) = P_1 : \ldots : P_n = R(a_1 : \ldots : a_n),$$

where $P_i = Ra_i$ for $1 \leq i \leq n$.

Let $P_i \in X_{\beta_i}$ and let $\bar{\beta} = \max\{\beta_i : 1 \leq i \leq n\}$. Clearly, $\bar{\beta} < \alpha$. On reindexing if necessary, we may assume that there exists an integer k, $1 \leq k \leq n$, such that $P_1, \ldots, P_k \in X_{\bar{\beta}}$ and $P_{k+1}, \ldots, P_n \notin X_{\bar{\beta}}$. We claim that $a \notin P_i^{[\omega^1]}$ for $i = 1,2,\ldots,k$. If not, let $a \in P_i^{[\omega^1]} = Q$ for some i, $1 \leq i \leq k$. By theorem 1.1, Q is a prime ideal of R; so, $a \in Q$ implies $P(Rb) \subseteq Q$ i.e., $P_1 : \ldots : P_n \subseteq Q$. Since Q is prime, $P_j \subseteq Q$ for some j, $1 \leq j \leq n$. Thus $P_j \subseteq Q \subseteq P_i$. This is possible only if $i = j$ so $P_i = Q = P_i^{[\omega^1]}$. Since P_i is non-zero, the last equation is impossible by theorem 1.1. Our claim is thus established. Since $a \in P_i$ but $a \notin P_i^{[\omega^1]}$ for $1 \leq i \leq k$, it follows that there exists an integer $m_i \in Z^+$ such that $a \in P_i^{m_i}$ but $a \notin P_i^{m_i+1}$ for $i = 1,2,\ldots,k$. Using 1.14, we have

$$a \in P_1^{m_1} \cap \ldots \cap P_k^{m_k} = R(a_1^{m_1} : \ldots : a_k^{m_k}).$$

Let $a = a' a_1^{m_1} : \ldots : a_k^{m_k}$. It is easily seen that $a' \notin P_i$, $1 \leq i \leq k$. Suppose $a' \in P$ for some $P \in X_{\bar{\beta}}$. Then $a \in P$ so $P(Rb) = P_1 : \ldots : P_n \subseteq P$. Consequently, $P_j \subseteq P$ for some j, $1 \leq j \leq n$. If $k < j$ then we get a contradiction with our choice of k; so $1 \leq j \leq k$. Since there are no inclusion relations among prime ideals belonging to $X_{\bar{\beta}}$, we have $P = P_j$. However, in this case, $a' \in P = P_j$ contradicts $a' \notin P_i$ for $1 \leq i \leq k$. Hence $a' \notin \cup\{P \mid P \in X_{\bar{\beta}}\}$. By the $\bar{\beta}^{th}$ transfinite induction hypothesis, Ra' is a product of a finite number of prime ideals of R. It follows that Ra is a product of a finite number of prime ideals of R. This completes the transfinite induction.

Using theorem 1.1, we have

$$R \setminus \{0\} = \bigcup_{\beta < \tau} \{R \setminus (\cup\{P \mid P \in X_\beta\})\};$$

it follows that every left ideal of R is two-sided, principally generated as a left ideal and that every left ideal of R is a product of a finite number of prime ideals of R. If $r_1 r_2 = 0$ then $Rr_1 Rr_2 = (0)$; since R is a prime ring, either $r_1 = 0$ or $r_2 = 0$. Hence R is a pli-domain. The remaining assertion now follows from theorem 2.1. This completes the proof.

§3. LOCAL FULLY LEFT GOLDIE IPLI-RINGS.

A ring R is called a local ring if R has a unique maximal left ideal, denoted by $J(R)$. As is well-known, $J(R)$ is a two-sided ideal of R and coincides with the set of all non-units of R. cf. Lambek [1, page 75]. The ideal theory developed in §§1 and 2 is clearly applicable to local fully left Goldie ipli-rings. In this section, we shall obtain further details about the ideal theory of these rings.

We recall some facts concerning ordinals. An ordinal ξ is called a prime component if ξ cannot be written as a sum of two ordinals strictly less than ξ. It is known that a non-zero ordinal ξ is a prime component if and only if $\xi = \omega^{\eta}$ for some uniquely determined ordinal η. It ic clear that the set $\{\lambda : \lambda < \omega^{\eta}\}$ is a (non-commutative) monoid with the usual addition.

Let α be a non-zero ordinal. Then α can be uniquely expressed as

$$\alpha = \omega^{\alpha_1} m_1 + \dots + \omega^{\alpha_k} m_k,$$

where k is a uniquely determined positive integer, $\alpha_1 > \dots > \alpha_k \geq 0$ are uniquely determined ordinals and m_i, $1 \leq i \leq k$, are uniquely determined positive integers; this expression for α is called the Cantor normal form of α; ω^{α_1} is called the initial prime component of α.

Suppose

$$\alpha = \omega^{\alpha_1} m_1 + \dots + \omega^{\alpha_k} m_k,$$
$$\beta = \omega^{\beta_1} n_1 + \dots + \omega^{\beta_\ell} n_\ell$$

are normal forms of non-zero ordinals α and β. The normal form of $\alpha + \beta$ is obtained as follows: If $\alpha_1 < \beta_1$ then $\alpha + \beta = \beta$; so, the normal forms of $\alpha + \beta$ and β are identical. Otherwise, there exist subscripts $i \leq k$ such that $\alpha_i \geq \beta_1$; let j denote the last such subscript. If $\alpha_j > \beta_1$ then the normal form of $\alpha + \beta$ is

$$\omega^{\alpha_1}m_1 + \ldots + \omega^{\alpha_j}m_j + \omega^{\beta_1}n_1 + \ldots + \omega^{\beta_\ell}n_\ell .$$

If $\alpha_j = \beta_1$ then the normal form of $\alpha + \beta$ is

$$\omega^{\alpha_1}m_1 + \ldots + \omega^{\alpha_{j-1}}m_{j-1} + \omega^{\alpha_j}(m_j + n_1) + \omega^{\beta_2}n_2 + \ldots \omega^{\beta_\ell}n_\ell .$$

These facts concerning ordinals may be found in Serpinski [1].

We now state the main results of this section.

3.1. THEOREM. Let R be a local semi-prime fully left Goldie ipli-ring. Then

(1) R is a pli-domain.

(2) There exists an ordinal τ such that $J^{[\omega^\tau]} = (0)$, where $J = J(R)$.

Let α be the first ordinal such that $J^{[\omega^\alpha]} = (0)$. For every $\beta < \alpha$, choose $x_\beta \in R$ such that $J^{[\omega^\beta]} = Rx_\beta$. Then

(3) Every non-zero $r \in R$ can be uniquely expressed as

$$r = u\, x_{\beta_1}^{m_1} : \ldots : x_{\beta_s}^{m_s} ,$$

where s is a non-negative integer, $m_i \in Z^+$ for $1 \leq i \leq s$, $0 \leq \beta_1 < \ldots < \beta_s < \alpha$ and u is a unit in R.

(4) Every left ideal of R is two-sided. The ideals of R form a well-ordered chain under reverse inclusion. The set of non-zero ideals of R forms a monoid under multiplication. Every non-zero ideal A of R can be uniquely expressed as

$$A = (Rx_{\beta_1})^{m_1} : \ldots : (Rx_{\beta_s})^{m_s},$$

where s is a non-negative integer, $m_i \in Z^+$ for $1 \leq i \leq s$ and $0 \leq \beta_1 < \ldots < \beta_s < \alpha$. Further,

$$A \to \omega^{\beta_s}m_s + \ldots + \omega^{\beta_2}m_2 + \omega^{\beta_1}m_1$$

is an order-preserving bijection of the set of all non-zero ideals of R under reverse inclusion with the set of all ordinals $< \omega^\alpha$. The same bijection is an anti-isomorphism of the monoid of all non-zero ideals of R under multiplication with the monoid of all ordinals $< \omega^{\alpha}$ under the usual addition.

(5) Every ideal of R is of the form J^{λ}, $\lambda \leq \omega^{\alpha}$, and $J^{[\omega^{\beta}]} = J^{\omega^{\beta}}$ for $\beta \leq \alpha$. In fact, if

$$\lambda = \omega^{\beta_s} m_s + \dots + \omega^{\beta_1} m_1$$

is the normal form of λ then

$$J^{\lambda} = (Rx_{\beta_1})^{m_1} : \dots : (Rx_{\beta_s})^{m_s}.$$

Recall that a ring R is called a <u>completely primary</u> ring if R is a primary ring and R/P(R) is a domain.

<u>3.2. THEOREM</u>. Let R be a local fully left Goldie ipli-ring. Then R is a completely primary pli-ring, every left ideal of R is two-sided and ideals of R form a well-ordered chain under reverse inclusion.

R is a domain if and only if the chain of all ideals of R under reverse inclusion is order isomorphic with the chain of all ordinals $\leq \omega^{\alpha}$ for some ordinal α.

<u>3.3. Proof of Theorem 3.1</u>. Let R be a local semi-prime fully left Goldie ipli-ring. Since J(R) is the set of all non-units of R, it follows from Johnson's first theorem that R is a prime ring. Since R/J(R) is a skew field, J(R) is a prime ideal of R. By Theorem 1.1, there exists an ordinal τ such that $J^{[\omega^{\tau}]} = (0)$. This proves (2). If α is the first ordinal with $J^{[\omega^{\alpha}]} = (0)$ then, by theorem 1.1, $\{J^{[\omega^{\beta}]} : \beta < \alpha\}$ is the set of all non-zero prime ideals of R. Theorem 2.2 now proves (1) and (3) and shows that every left ideal of R is two-sided.

Let A be a non-zero proper ideal of R. Using theorem 2.1 and 2.2, we have

$$A = (Rx_{\beta_1})^{m_1} : \dots : (Rx_{\beta_s})^{m_s}$$

where $s \geq 0$, $m_i \in Z^+$ for $1 \leq i \leq s$ and $0 \leq \beta_1 < \dots < \beta_s < \alpha$. Since there is an inclusion relation among any two distinct prime ideals of R, no non-identity permutation is permissible in the above prime factorization of A. Hence the above prime factorization of A is unique. It is now clear that the function defined by

$$A \to \omega^{\beta_s} m_s + \dots + \omega^{\beta_1} m_1,$$

where $A = (Rx_{\beta_1})^{m_1} : \dots : (Rx_{\beta_s})^{m_s}$ with $s \geq 0$, $m_i \in Z^+$ for $1 \leq i \leq s$ and $0 \leq \beta_1 < \dots < \beta_s < \alpha$, is a bijection of the set of all non-zero ideals of R with the set of all ordinals $< \omega^{\alpha}$.

Let $A = (Rx_{\beta_1})^{m_1} : \dots : (Rx_{\beta_s})^{m_s}$ and $B = (Rx_{\gamma_1})^{n_1} : \dots : (Rx_{\gamma_t})^{n_t}$ be non-zero ideals of R where s, t are non-negative integers, m_i and $n_j \in Z^+$ for $1 \leq i \leq s$ and $1 \leq j \leq t$, $0 \leq \beta_1 < \dots < \beta_s$ and $0 \leq \gamma_1 < \dots < \gamma_t$. Assume that

$$\omega^{\beta_s} m_s + \dots + \omega^{\beta_1} m_1 < \omega^{\gamma_t} n_t + \dots + \omega^{\gamma_1} n_1.$$

It follows that $\omega^{\beta_s} \leq \omega^{\gamma_t}$. Suppose $\omega^{\beta_s} < \omega^{\gamma_t}$. Then $Rx_{\gamma_t} \subset Rx_{\beta_s}$. Since R is a prime ring and $Rx_{\beta_s} \neq (0)$, it follows from theorem 1.1 that $Rx_{\gamma_t} \subset Rx_{\beta_s}^{m_s+1}$. Now, since $Rx_{\beta_s} \subset Rx_{\beta_i}$ for $1 \leq i \leq s-1$, theorem 1.5 yields

$$Rx_{\beta_s} A = (Rx_{\beta_s})^{m_s+1}.$$

Hence $A \supseteq (Rx_{\beta_s})A \supseteq (Rx_{\beta_s})^{m_s+1} \supset B$. Suppose $\omega^{\beta_s} = \omega^{\gamma_t}$; so, $x_{\beta_s} = x_{\gamma_t}$. Also

$$\omega^{\beta_s}(m_s-1) + \omega^{\beta_{s-1}} m_{s-1} + \dots + \omega^{\beta_1} m_1 < \omega^{\gamma_t}(n_t-1) + \omega^{\gamma_{t-1}} n_{t-1} + \dots + \omega^{\gamma_1} m_1.$$

An induction on the sum $\sum_{i=1}^{s} m_i + \sum_{j=1}^{t} n_j$ is thus available to conclude that the set of non-zero ideals of R is totally ordered under inclusion and that

$$A = (Rx_{\beta_1})^{m_1} : \ldots : (Rx_{\beta_s})^{m_s} \to \omega^{\beta_s} m_s + \ldots + \omega^{\beta_1} m_1$$

is an order isomorphism of the set of non-zero ideals of R under reverse inclusion and the set of all ordinals $< \omega^{\alpha}$. Since R has the ascending chain condition on left ideals, it follows that the set of all non-zero ideals of R is well-ordered under reverse inclusion. The assertion about the monoid of non-zero ideals of R easily follows from theorem 1.5. This proves (4).

It remains to prove (5). Firstly, we shall show that each ideal of R can be uniquely expressed as J^{λ} for some $\lambda \leq \omega^{\alpha}$. Since the set of all ideals of R under reverse inclusion is a well-ordered set and isomorphic with the set of all ordinals $\leq \omega^{\alpha}$, we can index the set of all ideals of R as $\{A_{\lambda} : \lambda \leq \omega^{\alpha}\}$ such that $A_{\lambda} \subset A_{\mu}$ if and only if $\mu < \lambda \leq \omega^{\alpha}$.

Trivially, $A_0 = J^0 = R$ and $A_1 = J^1 = J$. We proceed by a transfinite induction to show that $A_{\lambda} = J^{\lambda}$ for all $\lambda \leq \omega^{\alpha}$. Suppose $1 < \lambda \leq \omega^{\alpha}$ and suppose $A_{\mu} = J^{\mu}$ for all $\mu < \lambda$. If λ is a limit ordinal then, since A_{λ} is the first ideal properly contained in each A_{μ} for $\mu < \lambda$, we have

$$A_{\lambda} = \bigcap_{\mu<\lambda} A_{\mu} = \bigcap_{\mu<\lambda} J^{\mu} = J^{\lambda}.$$

Suppose λ is a non-limit ordinal, say $\lambda = \gamma + 1$. Then $A_{\gamma} \neq (0)$ and

$$J^{\lambda} = J^{\gamma+1} = J \cdot J^{\gamma} = JA_{\gamma} \subseteq A_{\gamma}.$$

Now observe that ${}_R A_{\gamma}$ and ${}_R R$ are isomorphic left R-modules. If $JA_{\gamma} = A_{\gamma}$ then it follows that $JR = J = R$, a contradiction. Thus $JA_{\gamma} \subset A_{\gamma}$. Also, if JA_{γ} is not a maximal submodule of A_{γ} then J is not a maximal submodule of ${}_R R$, a contradiction again. Hence $JA_{\gamma} = A_{\gamma+1}$; so, $J^{\lambda} = A_{\lambda}$. This completes the transfinite induction.

Now, if the cantor normal form of λ is

$$\lambda = \omega^{\beta_s} m_s + \ldots + \omega^{\beta_1} m_1$$

then, using (4), we have

$$A_\lambda = (Rx_{\beta_1})^{m_1} : \ldots : (Rx_{\beta_s})^{m_s}.$$

This proves (5). The theorem is now proved.

3.4. Proof of Theorem 3.2. Let P denote the prime radical of R. Since $P \subseteq J(R)$ and R/P is a semi-prime local fully left Goldie ipli-ring, by theorem 3.1, R/P is a pli-domain. Johnson's second theorem now shows that R is a completely primary ring.

Let $r \in R$, $r \notin P$. By theorem 3.1, every left ideal of R/P is two-sided; it follows that $Rr + P$ is a two-sided ideal of R. Let $Rr + P = Rr_0$ and let $r = r_1 r_0$, $r_1 \in R$. Since R/P is a domain and $r + P$ and $r_0 + P$ generate the same non-zero left ideal of R/P, it follows that $r_1 + P$ is a unit in R/P. Since P is nilpotent, r_1 must be a unit in R. Thus, $P \subset Rr_0 = Rr$. We have thus shown that every non-nilpotent left ideal of R is two-sided and contains P. It follows from 3.1 that the non-nilpotent left ideals of R form a well-ordered chain under reverse inclusion which is order-isomorphic with the chain of all ordinals $< \omega^\alpha$ for some ordinal α.

Suppose $P \neq (0)$; say $P = Rw_0$. Let $k + 1$ be the index of nilpotency of P, so $k \in Z^+$. Then $P^n = Rw_0^n$. If $w \in Rw_0^n$, $w \notin Rw_0^{n+1}$ then $w = rw_0^n$ where $r \in R$ but $r \notin Rw_0$. As shown above, $Rw_0 \subset Rr$. So, $Rw_0^{n+1} \subset Rrw_0^n = Rw$. We have now shown that if ℓ is a left ideal of R such that $\ell \subseteq Rw_0^n$ but $\ell \nsubseteq Rw_0^{n+1}$ then $Rw_0^{n+1} \subset \ell$. Now observe that Rw_0^n/Rw_0^{n+1} can be canonically considered as a left module over $\bar{R} = R/P$ and as such it is a cyclic module. Since the non-zero left ideals of $\bar{R}$ form a well-ordered chain under reverse inclusion and is order-isomorphic with the chain of all ordinals $< \omega^\alpha$ for some α, it follows that the non-zero left $\bar{R}$-submodules of

Rw_0^n/Rw_0^{n+1} form a well-ordered chain under reverse inclusion which is order-isomorphic with the chain of all ordinals $< \xi_n$ for some $\xi_n \leq \omega^\alpha$. To sum up, the left ideals of R form a well-ordered chain under reverse inclusion which is order-isomorphic with the chain of all ordinals $< \xi$, where

$$\xi = \omega^\alpha + \xi_1 + \dots + \xi_k.$$

Since $k \geq 0$ and $0 < \xi_k \leq \omega^\alpha$ for $1 \leq n \leq k$, therefore ξ is not of the form ω^β for any ordinal β.

If $w \in Rw_0^n$ but $w \notin Rw_0^{n+1}$ then $w = rw_0^n$ where $r \in R$, $r \notin Rw_0$. We have already shown that Rr is an ideal. It follows that $Rr \cdot Rw_0^n = Rw$; so Rw is an ideal of R. Since every principal left ideal is two-sided therefore every left ideal is two-sided and so a principal left ideal. It is now easy to complete the proof.

References.

The results in the last three §§ are due to the author. They were partially announced in Jategaonkar [6]. The proof of lemma 1.9 given in the text was suggested by Professor R. E. Johnson.

In Jategaonkar [3,4], we have constructed local pli-domains with embedded primes.

§4. IDEAL THEORY - II.

In this section, we shall consider rings which are fully left and right Goldie, ipli-ipri-rings with special attention to their ideal theory. It turns out that the ideal theory of such rings is very uncomplicated.

4.1. THEOREM. A ring R is a fully left and right Goldie ipli-ipri-ring if and only if R is a direct sum of a finite number of ideals, each of which is a left and right primary, fully left and right Goldie ipli-ipri-ring.

Theorem 4.1 is an analogue of Johnson's first theorem and allows us to concentrate on left-right primary, fully left and right Goldie, ipli-ipri-rings. We shall have to consider prime rings and non-prime primary rings of the above type separately.

4.2. THEOREM. Let R be a left and right primary, fully left and right Goldie, ipli-ipri-ring with $P(R) \neq (0)$. Then $R/P(R)$ is a simple ring. The only ideals of R are those of the form $\{P(R)\}^n$, $0 \leq n \leq k + 1$, where $k + 1$ is the index of nilpotency of $P(R)$. Further, R has a left-right Artinian, left-right primary, two-sided quotient ring.

4.3. THEOREM. Let R be a prime, fully left and right Goldie, ipli-ipri-ring. Then every non-zero proper prime ideal is maximal. A proper non-zero ideal is primary if and only if it is a finite power of a maximal ideal. Every proper ideal A can be expressed as a product of maximal ideals; these maximal ideals and the exponents with which they appear in the prime factorization of A are uniquely determined by A upto order. Any two ideals of R commute.

4.4. Proof of theorem 4.1. Follows immediately from Johnson's first and second theorems.

We now turn to theorem 4.2. Until further notice, R is a left and right primary, fully left and right Goldie, ipli-ipri-ring with $P(R) \neq (0)$.

4.5. LEMMA. R satisfies the regularity condition.

Proof. Let $\sigma : R \to \bar{R} = R/P(R)$ be the canonical epimorphism. Let $k + 1$ be the index of nilpotency of $P(R)$; since $P(R) \neq (0)$, $k \in Z^+$. Let A be the subset of $\{P(R)\}^k$ defined as follows: $x \in A$ if there exists an element $c \in R$, depending on x, such that $\sigma(c)$ is regular in $\bar{R}$ and $cx = 0$. Let $x_1, x_2 \in A$ and let $c_1, c_2 \in R$ such that $\sigma(c_1), \sigma(c_2)$ are regular in $\bar{R}$ and $c_1x_1 = 0 = c_2x_2$. As shown in the proof of Goldie's second theorem, $\bar{R}$ has the left common multiple property. Thus, there exist $b_1, b_2, c \in R$ such that $\sigma(c)$ is regular in $\bar{R}$ and $\sigma(c) = \sigma(b_1)\sigma(c_1) = \sigma(b_2)\sigma(c_2)$; i.e., $c = b_1c_1 + w_1 = b_2c_2 + w_2$; where $w_1, w_2 \in P(R)$. Since $w_i\{P(R)\}^k = (0)$ for $i = 1,2$, we have $c(x_1 \pm x_2) = 0$. So, $x_1 \pm x_2 \in A$. Also, given $r \in R$, there exist $r', d \in R$ with $\sigma(d)$ regular in $\bar{R}$ such that $\sigma(d)\sigma(r) = \sigma(r')\sigma(c_1)$ i.e., $dr = r'c_1 + w'$ where $w' \in P(R)$. Then $drx_1 = r'c_1x_1 + w'x_1 = 0$; so, $rx_1 \in A$. It is now clear that A is a two-sided ideal of R. Suppose $A \neq (0)$. Since R is a ipri-ring, $A = zR$ for some non-zero $z \in R$. There exists $b \in R$ such that $\sigma(b)$ is regular in $\bar{R}$ and $bz = 0$. So, $b \notin P(R)$ and $(RbR)zR = 0$. However, this contradicts the right-handed version of Johnson's third theorem. Hence $A = (0)$. Taking the symmetry of our hypothesis into account, we have shown that if $c \in R$ with $\sigma(c)$ regular in $\bar{R}$ then $cx \neq 0$ and $xc \neq 0$ for every non-zero $x \in \{P(R)\}^k$.

We are now in a position to prove the regularity condition. Suppose $a, y \in R$, $y \neq 0$, $ay = 0$ and $\sigma(a)$ is regular in $\bar{R}$. Then $y \in P(R)$. Since $y \neq 0$, there exists n, $(1 \leq n \leq k)$, such that

$y \in \{P(R)\}^n$ but $y \notin \{P(R)\}^{n+1}$. Let $\psi: R \to R/\{P(R)\}^{n+1} = \tilde{R}$ be the canonical epimorphism. It is clear that the first part of the proof is applicable to $\tilde{R}$. This yields $\psi(y) = \psi(0)$ i.e., $y \in \{P(R)\}^{n+1}$, a contradiction. Thus, a is right regular in R. Similarly, a is left regular in R. This proves the lemma.

4.6. LEMMA. R has a two-sided quotient ring Q which is a ipli-ipri, left and right Artinian primary ring. Further, $P(R) = P(Q) \cap R$ and $P(Q) = QP(R) = P(R)Q$.

Proof. Lemma 4.5 and Small's theorem shows that R has a two-sided quotient ring Q which is a left and right Artinian ring. By lemma I.4.5, $Q/P(Q)$ is a two-sided quotient ring of $R/P(R)$. By Johnson's second theorem, $R/P(R)$ is a prime left Goldie ring so $Q/P(Q)$ is a simple Artinian ring and Q is a primary ring. If A is an ideal of Q then $A = (A \cap R)Q = Q(A \cap R)$; so Q is a ipli-ipri-ring. The remaining assertions are contained in lemma I.4.5.

4.7. LEMMA. Let Q be a primary left and right Artinian ring in which $P(Q) = Qz = z'Q$. Then $P(Q) = zQ = Qz'$.

Proof. Since our hypothesis is symmetric, using theorem II.4.12, there exists $w \in Q$ such that $P(Q) = Qw = wQ$. Thus, $wu = z$ and $vz = w$ for some $u, v \in Q$; so, $vwu = w$. Since $vw \in Qw = wQ$, there exists $v_1 \in Q$ such that $vw = wv_1$. Thus, $w = vwu = wv_1u$ i.e., $Qw(1 - v_1u) = 0$. Since Q is primary, $\ell_Q(Qw) \subseteq P(Q)$ so $1 - v_1u \in P(Q)$. It follows that u is a unit. Thus $Qz = Qw = wQ = zQ$. Similarly, $Qz' = z'Q$. This proves the lemma.

4.8. LEMMA. If $P(R) = Rz = z'R$ then $P(R) = zR = Rz'$.

Proof. $Rz = z'R$ so $uz = z'$ and $z'v = z$ for some $u,v \in R$. Therefore $z = uzv$. Since $zv \in Rz$, we have $zv = v_1z$. Thus $(1 - uv_1)z = 0$. By 4.6, R has a two-sided left and right Artinian ipli-ipri quotient ring Q. Also, $P(Q) = QP(R) = QRz = Qz$. By 4.7, $Qz = zQ$. Now, $(1 - uv_1)z = 0$ gives $(1 - uv_1)P(Q) = (0)$. It follows that $(1 - uv_1) \in P(Q) \cap R = P(R)$. Hence u is a unit R. $uz = z'$ now yields $Rz = Rz' = P(R)$. Similarly, $P(R) = zR$. This proves the lemma.

4.9. Proof of theorem 4.2. Firstly we show that $R/P(R)$ is a simple ring. If not, R has a non-nilpotent proper ideal, say Ra. Let $P(R) = Rz$. Clearly, $RaRz = Raz \subseteq Rz$. If $Raz = Rz$ then $(ua - 1)zR = 0$ for some $u \in R$. By 4.8, $P(R) = zR$ so $(ua-1)P(R)=0$. Applying Johnson's third theorem from the right side, we obtain $(ua - 1) \in P(R)$. Hence a is a unit in R and $Ra = R$, contrary to our hypothesis. We thus have $Raz \subset Rz$. Let $\tilde{R} = R/Raz$ and let $\phi: R \to \tilde{R}$ be the canonical epimorphism. Thus $\phi(a)$ is a non-regular element of $\tilde{R}$. However, it is easy to see that this contradicts the regularity condition in $\tilde{R}$ established in 4.5. Hence $R/P(R)$ must be a simple ring.

Let B be a proper non-zero ideal of R. Then $B \subseteq P(R)$ since $R/P(R)$ is simple. Choose $n \in Z^+$ such that $B \subseteq \{P(R)\}^n$ but $B \nsubseteq \{P(R)\}^{n+1}$. Clearly, $\{P(R)\}^n = Rz^n$. Let $X = \{x \in R \mid xz^n \in B\}$. X is a left ideal of R and $Xz^n = B$. By 4.8, $P(R) = zR$ so that $\{P(R)\}^n = z^nR$. Since B is an ideal of R, we have $B = BR = Xz^nR = XRz^n$, which shows that X is an ideal of R. Since $R/P(R)$ is simple, either $X = R$ or $X \subseteq Rz$. In the later case, we get $B = Xz^n \subseteq Rz^{n+1}$, a contradiction. Hence $X = R$ and $B = Rz^n = \{P(R)\}^n$. The assertion about the quotient ring is proved in lemma 4.6. This completes the proof.

We now abandon the restrictions on R.

4.10. Proof of theorem 4.3. Let Ra be a prime ideal of R. Then $\tilde{R} = R/Ra^2$ is a left-right primary, left-right fully Goldie ipli-ipri-ring with non-zero prime radical. By theorem 4.2, $\tilde{R}/P(\tilde{R}) \cong R/Ra$ is a simple ring. Hence Ra is a maximal ideal of R.

Let Rb be a primary ideal of R and let $P(Rb) = Rc$. By theorem 1.3, Rc is a prime ideal of R so a maximal ideal of R. Further, there exists $n \in Z^+$ such that $Rc^n \subset Rb \subseteq Rc$. Applying theorem 4.2 to R/Rc^n, we conclude that $Rb = Rc^m$ for some $m \in Z^+$. The rest follows from theorem 1.4. This completes the proof.

We end this section with a proposition which gives some examples of rings, other than pli-pri-rings, to which the results of this section are applicable.

4.11. PROPOSITION. Let S be a simple left and right Noetherian ring and let ρ be an automorphism of S. Then $R = S[x,\rho]$ is a left and right Noetherian prime ipli-ipri-ring.

Proof. Let I be a non-zero ideal of R and let

$$f(x) = a_n x^n + \dots + a_0$$

be a polynomial of least degree in I, $a_n \neq 0$. Since $Sa_nS = S$, there exist $s_i, t_i \in S$ such that $1 = \sum_{i=1}^{m} s_i a_n t_i$. Since ρ is an automorphism, there exist t_i', $1 \leq i \leq m$, such that $t_i = \rho^n(t_i')$. Then $g(x) = \sum_{i=1}^{m} s_i f(x) t_i' = x^n + b_{n-1}x^{n-1} + \dots + b_0 \in I$. It is easily seen that $I = Rg(x)$. It thus follows that R is a prime ipli-ipri-ring. A straightforward adoption of the Hilbert basis theorem shows that R is a left and right Noetherian ring. This completes the proof.

It is clear that Rx is an ideal of $R = S[x,\rho]$. Thus R/Rx^k, $k > 1$, is a left-right primary, left-right Noetherian, ipli-ipri-ring with non-zero prime radical.

References

The results of this section are due to Robson [3]. Some of our proofs are new.

CHAPTER IV

PLI-DOMAINS

INTRODUCTION

In this chapter, we briefly consider some aspects of pli-domains.

As is well-known, Euclidean domains are easier to handle than arbitrary PID's. However, the structure of Euclidean domains is unknown. In §1, we define a transfinite left division algorithm which generalizes the Euclidean algorithm on the polynomial rings in one indeterminate over fields and determine the structure of rings which have such an algorithm. These rings are iterated skew polynomial rings of a special type over skew fields. In §2 we show that embedded primes yield a lower bound for the right global dimension of pli-domains.

§1. RINGS WITH TRANSFINITE LEFT DIVISION ALGORITHM.

Let R be a non-zero ring (not necessarily with 1). Let d be an ordinal-valued function defined on $R\setminus\{0\}$. Let $-\infty < \eta$ and $(-\infty) + \eta = \eta + (-\infty) = (-\infty) + (-\infty) = -\infty$ for every ordinal η. Put $d(0) = -\infty$. d is called a transfinite left division algorithm on R if the following conditions hold for every $a, b \in R$:

(D1) $d(a-b) \leq \max\{d(a), d(b)\}$.

(D2) $d(ab) = d(b) + d(a)$

(D3) If $b \neq 0$ then there exist $q, r \in R$ such that $d(r) < d(b)$ and

$$a = qb + r.$$

It may be worthwhile to point out that our definition of 'transfinite left division algorithm' is not a generalization of the usual definition of an Euclidean algorithm on commutative domains. Indeed our definition is so formulated that the only commutative rings with a transfinite left division algorithm are fields and polymonial rings in one indeterminate over fields.

Notice that if a ring R with 1 has a transfinite left division algorithm then D2 and D3 show that R is a pli-domain. In this section, we shall determine the structure of such domains.

A definition is needed. Let D be a domain and $\rho: D \to D$ a monomorphism. A mapping $\delta: D \to D$ is called a ρ-derivation on D if

$$\delta(a+b) = \delta(a) + \delta(b)$$
$$\delta(ab) = \rho(a)\delta(b) + \delta(a)b$$

for all $a, b \in D$. Let $D[x, \rho, \delta]$ be the set of all formal polynomials in x with coefficients in D written on the left of powers of x. Define equality and addition as usual; define multiplication by assuming the distributive laws and the rule

$$xa = \rho(a)x + \delta(a)$$

for all $a \in D$. It is easy to check that $D[x, \rho, \delta]$ is a ring.

Suppose R is a domain, α is an ordinal number and $\{R_\beta : \beta < \alpha\}$ is a chain of subdomains of R such that $R = \bigcup_{\beta<\alpha} R_\beta$ and, if $0 < \beta < \alpha$ then $R_\beta = (\bigcup_{\gamma<\beta} R_\gamma)[x_\beta, \rho_\beta, \delta_\beta]$. Then R is called a generalized skew polynomial extension of R_0 and is denoted by $R = R_0[x_\beta, \rho_\beta, \delta_\beta : 0 < \beta < \alpha]$.

We now state the main result of this section.

1.1. THEOREM. Let R be a non-zero ring. R has a transfinite left division algorithm if and only if

$$R = K[x_\beta, \rho_\beta, \delta_\beta : \ 0 < \beta < \alpha]$$

where α is an ordinal, K is a subskew field of R and ρ_β is a monomorphism into K for $0 < \beta < \alpha$.

1.2. LEMMA. Let K be a skew field and let

$$R = K[x_\beta, \rho_\beta, \delta_\beta : \ 0 < \beta < \alpha]$$

where

$$\rho_\beta : \ K[x_\gamma, \rho_\gamma, \delta_\gamma : \ 0 < \gamma < \beta] \to K$$

is a monomorphism for $0 < \beta < \alpha$ and δ_β is a ρ_β-derivation on $K[x_\gamma, \rho_\gamma, \delta_\gamma : \ 0 < \gamma < \beta]$. Then R has a transfinite left division algorithm.

Proof. Let $\overline{R}_1 = K$ and $\overline{R}_\beta = K[x_\gamma, \rho_\gamma, \delta_\gamma : \ 0 < \gamma < \beta]$ for $0 < \beta \leq \alpha$; so, $R = \overline{R}_\alpha$. Let $\mathcal{A}$ be the set of all pairs (S,d), where $S = \overline{R}_\beta$ for some $\beta \leq \alpha$ and d is a transfinite left division algorithm on $S = \overline{R}_\beta$ such that $(\text{range } d) \subseteq \{\eta : \eta < \omega^\beta\}$. If (S,d) and (S',d') are in $\mathcal{A}$, we define $(S,d) \leq (S',d')$ if S is a subdomain of S' and $d = d'|S$. It is clear that $\mathcal{A}$ is partially ordered by $\leq$.

Since the function on K assigning $-\infty$ to 0 and 0 to all non-zero elements of K is a transfinite left division algorithm on K, the set $\mathbf{A}$ is non-empty. It is easily seen that $(\mathbf{A}, \leq)$ is an inductive set. By Zorn's lemma, $\mathbf{A}$ has a maximal element, say $(\overline{R}_\beta, d)$. We claim that $\overline{R}_\beta = R$. If not, consider the domain $S = \overline{R}_{\beta+1} = \overline{R}_\beta[x_\beta, \rho_\beta, \delta_\beta]$. Every non-zero $s \in S$ can be uniquely expressed as

$$s = \sum_{i=0}^{n} r_i\, x_\beta^i ,$$

where n is a non-negative integer, $r_i \in \overline{R}_\beta$ for $0 \leq i \leq n$ and $r_n \neq 0$. Put

$$f(s) = \omega^\beta n + d(r_n) \qquad \text{if } s \text{ is as above;}$$

$$f(0) = -\infty.$$

Clearly, f is a well-defined function on S and (range f) $\subseteq \{\eta : \eta < \omega^{\beta+1}\}$ since (range d) $\subseteq \{\eta : \eta < \omega^\beta\}$. Further f extends d. We shall show that f is a transfinite left division algorithm on S.

It is immediate that D1 holds for f since it holds for d.

Let a, b be non-zero elements of S, say

$$a = \sum_{i=0}^{m} r_i\, x_\beta^i \text{ and } b = \sum_{j=0}^{n} r_i'\, x_\beta^j$$

where m, n are non-negative integers, $r_i, r_j' \in \overline{R}_\beta$ for $0 \leq i \leq m$, $0 \leq j \leq n$ and $r_m \neq 0$, $r_n' \neq 0$. An easy computation shows that

$$ab = r_m\, \rho_\beta^m (r_n')\, x_\beta^{m+n} + \sum_{i=0}^{m+n-1} r_i''\, x_\beta^i,$$

where $r_i'' \in \overline{R}_\beta$ for $0 \leq i \leq m+n-1$. If $m = 0$, then

$$ab = \sum_{j=0}^{n} r_o\, r_j'\, x_\beta^j ;$$

so,

$$f(ab) = \omega^{\beta}n + d(r_o r_n') = \omega^{\beta}n + d(r_n') + d(r_o)$$

$$= f(b) + f(a).$$

If $m > 0$, then $\rho_{\beta}^{m}(r_n') \in K^*$ so that $d(\rho_{\beta}^{m}(r_n')) = 0$.

Then

$$f(ab) = \omega^{\beta}(n+m) + d(r_m\, \rho_{\beta}^{m}(r_n'))$$

$$= \omega^{\beta}n + \omega^{\beta}m + d(\rho_{\beta}^{m}(r_n')) + d(r_m)$$

$$= \omega^{\beta}n + d(r_n') + \omega^{\beta}m + d(r_m)$$

$$= f(b) + f(a),$$

since $m > 0$ and $d(r_n') < \omega^{\beta}$ implies $d(r_n') + \omega^{\beta}m = \omega^{\beta}m$. Hence D2 holds for f on S.

It remains to show that D3 holds for f. Let a, b be as stated above with $b \neq 0$. If $f(a) < f(b)$ then we take $q = 0$ and $r = a$. Assume that $f(a) \geq f(b)$ i.e., $\omega^{\beta}m + d(r_m) \geq \omega^{\beta}n + d(r_n')$ so that $m \geq n$ and if $m = n$ then $d(r_m) \geq d(r_n')$. Suppose $m > n$. Then we have

$$a - \{r_m[\rho_{\beta}^{m-n}(r_n')]^{-1}\, x_{\beta}^{m-n}\, b\} = \sum_{i=0}^{m-1} t_i\, x_{\beta}^{i}$$

where $t_i \in \overline{R}_{\beta}$ for $0 \leq i \leq m-1$. In a finite number of steps of this kind, we get $q_1 \in S$ such that

$$a - q_1 b = \sum_{i=0}^{n} t_i'\, x_{\beta}^{i},$$

where $t_i' \in \overline{R}_{\beta}$ for $0 \leq i \leq n$. If $d(t_n) < d(r_n')$ then we can take $q = q_1$ and $r = \sum_{i=0}^{n} t_i'\, x_{\beta}^{i}$ so $f(r) < f(b)$. It thus suffices to consider the case when $m = n$ and $d(r_n) \geq d(r_n')$. Since d is a transfinite left division algorithm on $\overline{R}_{\beta}$ and $r_n, r_n' \in \overline{R}_{\beta}$, we have $q_o, z \in \overline{R}_{\beta}$ such that $d(z) < d(r_n')$ and $r_n = q_o r_n' + z$. It follows that

$$a - q_o b = z\, x_{\beta}^{n} + \sum_{i=0}^{n-1} z_i\, x_{\beta}^{i}$$

where $z_i \in \overline{R}_\beta$ for $0 \leq i \leq n-1$. Clearly,

$$f(a-q_0b) = \omega^\beta n + d(z) \leq \omega^\beta n + d(r_n') = f(b).$$

Hence D3 holds for f on S.

Summing up, $(S, f) \in \mathcal{A}$. However, this contradicts the maximality of $(\overline{R}_\beta, d)$ in $(\mathcal{A}, \leq)$. Hence $\overline{R}_\beta = R$. This proves our claim and concludes the proof.

1.3. LEMMA. Let R be a non-zero ring with a transfinite left division algorithm d. Then there exists a subskew field K of R and an ordianl α such that

$$R = K[x_\beta, \rho_\beta, \delta_\beta: \ 0 < \beta < \alpha]$$

where

$$\rho_\beta\,(K[x_\gamma, \rho_\gamma, \delta_\gamma: \ 0 < \gamma < \beta]) \subseteq K$$

for every $0 < \beta < \alpha$.

Proof. Let a be a non-zero element of least degree in R. By D3, there exists a non-zero $e \in R$ such that $a = ea$ so $(e^2 - e)a = 0$. By D2, R has no zero-divisors. Thus $e^2 = e$. Now, for every $r \in R$, $e(er-r) = (re-r)e = 0$ shows that $r = re = er$. Hence e is the unity of R. Also, $d(e) = d(e^2) = d(e) + d(e)$ implies $d(e) = 0$; since $0 = d(e) = d((-e)(-e)) = d(-e) + d(-e)$, it follows that $d(-e) = 0$. So, $d(-r) = d((-e)(r)) = d(r) + d(-e) = d(r)$ for every $r \in R$. Consequently, we have

(D4): For $a, b \in R$, $d(a+b) \leq \max\{d(a), d(b)\}$.

Suppose $d(a) > d(b)$. Then

$$d(a) = d(a+b-b) \leq \max\{d(a+b), d(b)\} \leq d(a).$$

Hence

(D5): If $d(a) > d(b)$ then $d(a+b) = d(a)$.

Let $K = \{r \in R | d(r) \leq 0\}$. (D1) and (D2) show that K is a unitary subring of R. By (D3), for every non-zero $a \in K$, there exists $b \in R$ with $e = ba$; so, $0 = d(e) = d(a) + d(b) = d(b)$; so, $b \in K$. It follows that K is a subskew field of R.

We now show that there exists an ordinal α and a chain $\{\mathfrak{C}_\lambda : 0 < \lambda \leq \alpha\}$ of chains of subdomains of R satisfying the following conditions:

1) For every λ, $1 \leq \lambda \leq \alpha$, $\mathfrak{C}_\lambda = \{R_\mu : \mu < \lambda\}$ where $R_0 = K$ and for $0 < \mu < \lambda$,

$$R_\mu = (\bigcup_{\nu < \mu} R_\nu)[x_\mu, \rho_\mu, \delta_\mu]$$

with

$$\rho_\mu (\bigcup_{\nu < \mu} R_\nu) \subseteq K;$$

Further, if $r \in R$, $s \in R_\mu$ and $d(r) \leq d(s)$ then $r \in R_\mu$.

2) If $\lambda_1 < \lambda_2$ then $\mathfrak{C}_{\lambda_1}$ is an initial segment of $\mathfrak{C}_{\lambda_2}$.

3) $R = \bigcup\{R_\mu : R_\mu \in \mathfrak{C}_\alpha\}$.

It is evident that the existence of $\mathfrak{C}_\alpha$ will prove the lemma.

Put $\mathfrak{C}_1 = \{K\}$. We now proceed by a transfinite induction. Suppose $\mathfrak{C}_\lambda$ is defined for each $\lambda < \lambda_0$ and that conditions 1) and 2) stated above hold for $\{\mathfrak{C}_\lambda : 0 < \lambda < \lambda_0\}$. We want to get a chain $\mathfrak{C}_{\lambda_0}$ of subdomains of R such that the conditions 1) and 2) hold for $\{\mathfrak{C}_\lambda : 0 < \lambda \leq \lambda_0\}$.

If λ_0 is a limit ordinal then $\mathfrak{C}_{\lambda_0} = \bigcup_{\lambda < \lambda_0} \mathfrak{C}_\lambda$ works. Assume that λ_0 is a non-limit ordinal, say $\lambda_0 = \gamma+1$ and that $\bigcup_{\mu < \gamma} R_\mu \neq R$. We shall obtain a subdomain R_γ of R satisfying the following conditions:

(i) $R_\gamma = (\bigcup_{\mu < \gamma} R_\mu)[x_\gamma, \rho_\gamma, \delta_\gamma]$

where

$$\rho_\gamma (\bigcup_{\mu < \gamma} R_\mu) \subseteq K.$$

(ii) If $r \in R$, $s \in R_\gamma$ and $d(r) \leq d(s)$ then $r \in R_\gamma$.
Once such a domain is obtained, we can put $\mathfrak{S}_{\lambda_0} = \{R_\lambda : 0 < \lambda < \lambda_0\}$ to complete the transfinite induction. A cardinality argument then proves the existence of the required ordinal α.

Let x_γ be an element of least degree in the set $R \setminus (\bigcup_{\mu<\gamma} R_\mu)$. We claim that the initial prime component of $d(x_\gamma)$ must be strictly greater than the initial prime component of $d(a)$ for every non-zero $a \in \bigcup_{\mu<\gamma} R_\mu$. It is clear that the initial prime component of $d(x_\gamma)$ cannot precede the initial prime component of $d(a)$ for any non-zero $a \in \bigcup_{\mu<\gamma} R_\mu$. Suppose there exists a non-zero $a \in \bigcup_{\mu<\gamma} R_\mu$ such that $d(x_\gamma)$ and $d(a)$ have the same initial prime components. Let the normal forms of $d(x_\gamma)$ and $d(a)$ be

$$d(x_\gamma) = \omega^\beta m_1 + \omega^{\eta_2} m_2 + \dots + \omega^{\eta_k} m_k,$$

$$d(a) = \omega^\beta n_1 + \omega^{\xi_2} n_2 + \dots\dots \omega^{\xi_\ell} n_\ell$$

where $\beta > \eta_2 > \dots > \eta_k \geq 0$; $\beta > \xi_2 > \dots > \xi_\ell \geq 0$; k, ℓ, m_i, n_j are positive integers. Our choice of x_γ shows that $m_1 \geq n_1$. By (D3), there exist $q_1, r_1 \in R$ such that $d(r_1) < d(a)$ and $x_\gamma = q_1 a + r_1$. Clearly, $q_1 \neq 0$ so that $d(q_1 a) = d(a) + d(q_1) \geq d(a)$. Using (D5), we have

$$d(x_\gamma) = d(q_1 a + r_1) = d(q_1 a) = d(a) + d(q_1).$$

If $m_1 = n_1$ then $d(q_1) < d(a)$. It follows that $q_1, r_1, a \in \bigcup_{\mu<\gamma} R_\mu$ so that $x_\gamma \in \bigcup_{\mu<\gamma} R_\mu$, a contradiction. Thus, $m_1 > n_1$ and the normal form of $d(q_1)$ is

$$d(q_1) = \omega^\beta (m_1 - n_1) + \omega^{\zeta_2} p_2 + \dots + \omega^{\zeta_i} p_i$$

where $\beta > \zeta_2 > \dots > \zeta_i$, and i, p_j are positive integers. It follows that $d(q_1) < d(x_\gamma)$; so, $q_1, r_1, a \in \bigcup_{\mu<\gamma} R_\mu$ and $x_\gamma \in \bigcup_{\mu<\gamma} R_\mu$, a contradiction. This establishes our claim.

Now, let b be an arbitrary non-zero element of $\bigcup_{\mu < \gamma} R_\mu$. By (D3), there exist $q, r \in R$ such that $d(r) < d(x_\gamma)$ and $x_\gamma b = q\, x_\gamma + r$. Clearly, $r \in \bigcup_{\mu < \gamma} R_\mu$. Further, using the claim proved above, we have

$$d(x_\gamma) = d(b) + d(x_\gamma) = d(x_\gamma b).$$

Thus, $q \neq 0$ and using (D5), we have

$$d(x_\gamma) = d(x_\gamma b) = d(q\, x_\gamma + r)$$

$$= d(q\, x_\gamma) = d(x_\gamma) + d(q).$$

It follows that $d(q) = 0$ so $q \in K^*$. If $q_1, r_1 \in R$ are such that $d(r_1) < d(x_\gamma)$ and $x_\gamma b = q_1\, x_\gamma + r_1$ then $(q-q_1)\, x_\gamma = r_1 - r$; so,

$$d(x_\gamma) + d(q-q_1) = d(r_1-r) < d(x_\gamma).$$

This implies $d(q-q_1) = -\infty$ i.e. $q = q_1$ so that $r = r_1$. Hence q, r are uniquely defined by b. Let

$$\rho_\gamma : \bigcup_{\mu < \gamma} R_\mu \to K$$

be the function defined by $\rho_\gamma(b) = q$ if $b \neq 0$; $\rho_\gamma(0) = 0$. Let

$$\delta_\gamma : \bigcup_{\mu < \gamma} R_\mu \to \bigcup_{\mu < \gamma} R_\mu$$

defined by $\delta_\gamma(b) = r$ if $b \neq 0$ and $\delta_\gamma(0) = 0$. Thus

$$x_\gamma\, b = \rho_\gamma(b)\, x_\gamma + \delta_\gamma(b)$$

for every $b \in \bigcup_{\mu < \gamma} R_\mu$. It is now easy to see that ρ_γ is a monomorphism and δ_γ is a ρ_γ-derivation.

If possible, let $\sum_{i=0}^{n} r_i\, x_\gamma^i = 0$, where $r_i \in \bigcup_{\mu < \gamma} R_\mu$ and $r_n \neq 0$. By the claim proved earlier, $d(r_i\, x_\gamma^i) > d(r_j\, x_\gamma^j)$ if $i > j$ and $r_i \neq 0$. Using (D5) repeatedly, we obtain

$$-\infty = d(0) = d\left(\sum_{i=0}^{n} r_i\, x_\gamma^i\right) = d(x_\gamma^n) + d(r_n),$$

a contradiction. Hence, the subring of R generated by $\bigcup_{\mu<\gamma} R_\mu$ and x_γ is $(\bigcup_{\mu<\gamma} R_\mu)[x_\gamma, \rho_\gamma, \delta_\gamma]$; we denote this subring by R_γ.

Let $r \in R$ and $s \in R_\gamma$ with $d(r) \leq d(s)$. We shall show that $r \in R_\gamma$. If $r \in \bigcup_{\mu<\gamma} R_\mu$, there is nothing to prove. Assume that $r \notin \bigcup_{\mu<\gamma} R_\mu$. Then $s \in R_\gamma$ but $s \notin \bigcup_{\mu<\gamma} R_\mu$. It follows that the initial prime component of $d(s)$ is ω^β, the initial prime component of x_γ. Thus the initial prime component of $d(r)$ is $\leq \omega^\beta$. However, by the choice of x_γ, $d(r) \geq d(x_\gamma)$. It follows that the initial prime components of $d(r)$, $d(s)$ and $d(x_\gamma)$ are equal to ω^β. Let

$$d(x_\gamma) = \omega^\beta m_1 + \omega^{\eta_2} m_2 + \ldots + \omega^{\eta_k} m_k$$

$$d(r) = \omega^\beta n_1 + \omega^{\beta_2} n_2 + \ldots + \omega^{\beta_t} n_t$$

be the normal forms of $d(x_\gamma)$ and $d(r)$. Since $d(r) \geq d(x_\gamma)$, we have $n_1 \geq m_1$. By (D3), there exist $q', r' \in R$ such that $d(r') < d(x_\gamma)$ and $r = q'x_\gamma + r'$. It is clear that $r' \in \bigcup_{\mu<\gamma} R_\mu$ and

$$d(r) = d(q' x_\gamma + r') = d(q' x_\gamma)$$

$$= d(x_\gamma) + d(q').$$

It follows that $d(q') < d(r)$. A transfinite induction now suffices to conclude that $r \in R_\gamma$. We have thus shown that R_γ satisfies conditions (i) and (ii). As indicated earlier, this completes the proof of the lemma.

<u>1.4. Proof of theorem 1.1.</u> Follows from lemmas 1.2 and 1.3.

<u>References</u>

Jacobson [1] and Cohn [1] have considered rings with a left division algorithm with values in non-negative integers. The results of this section were announced in Jategaonkar [7]. For a construction of a ring with transfinite left division algorithm, see Jategaonkar [4].

§2. RIGHT GLOBAL DIMENSION.

In this section, we obtain a lower bound for the right global dimension of a fully left Goldie ipli-domain.

The main result of this section is the following.

2·1. THEOREM. Let R be a fully left Goldie ipli-domain. Let A be a proper prime ideal of R and let λ be an ordinal such that $A^{[\omega^\lambda]} \neq (0)$. Then

r. gl. dim $R = \infty$ if card $\lambda \geq \aleph_\omega$.

$\geq n+2$ if card $\lambda = \aleph_n$ where n is a non-negative integer.

≥ 2 if λ is a non-zero integer.

We need a lemma.

2·2. LEMMA. Let R be a prime fully left Goldie ipli-ring, A be a proper prime ideal of R and let λ be an ordinal such that $A^{[\omega^\lambda]} \neq (0)$. Then there exists a set of principal right ideals of R which forms a well-ordered chain under inclusion order-isomorphic with the chain of all ordinals $< \lambda$.

Proof. For $\beta \leq \lambda$, choose $x_\beta \in R$ such that $A^{[\omega^\beta]} = Rx_\beta$. By theorem III·1·1, each Rx_β is a prime ideal of R and $Rx_\beta \subset Rx_\gamma$ if $\gamma < \beta \leq \lambda$. By III·1·5, $Rx_\beta = Rx_\beta Rx_\gamma = Rx_\beta x_\gamma$ for $\gamma < \beta \leq \lambda$. If $x_\beta = ux_\beta x_\gamma$ and $x_\beta x_\gamma = wx_\beta$ then $(1-uw)x_\beta = 0$ $= (1-wu)x_\beta x_\gamma$. By Johnson's third theorem, x_β and x_γ are left regular elements in R . It follows that w is a unit in R .

Since Rx_β is an ideal of R , for every $r \in R$, there exists $r' \in R$ such that $x_\beta r = r'x_\beta$. By Johnson's third theorem, x_β is left regular in R so that r' is uniquely determined by r . Let $\rho_\beta \colon R \to R$ be the map defined by $r \mapsto r'$. Clearly, ρ_β is an endomorphism. Since $(Rx_\beta)(\ker \rho_\beta) = (0)$ and since R is a prime ring, it follows that $\ker \rho_\beta = (0)$ so ρ_β is a monomorphism. The first paragraph now shows that $\rho_\beta(x_\gamma)$ is a unit in R for every $\gamma < \beta \leq \lambda$.

Now, for fixed $\gamma < \beta < \lambda$, consider the element

$$a = [\rho_\lambda(x_\beta)]^{-1} x_\lambda [\rho_\beta(x_\gamma)]^{-1} x_\beta .$$

From what has been proved above, a is an element of R. Also,

$$a = [\rho_\lambda(x_\beta)]^{-1} \rho_\lambda ([\rho_\beta(x_\gamma)]^{-1} x_\beta) x_\lambda$$

$$= [\rho_\lambda(x_\beta)]^{-1} \rho_\lambda ([\rho_\beta(x_\gamma)]^{-1} x_\beta x_\gamma) [\rho_\lambda(x_\gamma)]^{-1} x_\lambda .$$

Since $x_\beta x_\gamma = \rho_\beta(x_\gamma) x_\beta$ and since $\rho_\beta(x_\gamma)$ is a unit in R, we get

$$a = [\rho_\lambda (x_\gamma)]^{-1} x_\lambda .$$

To sum up, we have shown that

$$[\rho_\lambda (x_\gamma)]^{-1} x_\lambda R \subseteq [\rho_\lambda (x_\beta)]^{-1} x_\lambda R .$$

If equality holds in the above inclusion relation, then there exists $r \in R$ such that

$$[\rho_\lambda (x_\beta)]^{-1} x_\lambda = [\rho_\lambda (x_\gamma)]^{-1} x_\lambda r .$$

Then

$$x_\lambda = \rho_\lambda (x_\beta) [\rho_\lambda (x_\gamma)]^{-1} x_\lambda r$$

$$= \rho_\lambda ([\rho_\beta(x_\gamma)]^{-1} x_\beta x_\gamma) [\rho_\lambda(x_\gamma)]^{-1} x_\lambda r$$

$$= x_\lambda [\rho_\beta (x_\gamma)]^{-1} x_\beta r .$$

Since R is a prime ring and Rx_λ is a non-zero ideal of R, x_λ is a right regular element of R. Cancelling x_λ, we get

$$1 = [\rho_\beta (x_\gamma)]^{-1} x_\beta r$$

$$= x_\beta r [\rho_\beta (x_\gamma)]^{-1}$$

$$= \rho_\beta(r[\rho_\beta (x_\gamma)]^{-1}) x_\beta .$$

Thus, x_β is a unit in R so $Rx_\beta = R$. Since $Rx_\beta \subseteq A$, we have $A = R$, contrary to our hypothesis. Hence

$$\{[\rho_\lambda (x_\gamma)]^{-1} x_\lambda R : \gamma < \lambda\}$$

is a well-ordered chain of right ideals of R under inclusion. This completes the proof.

We now state without proof a theorem of B.L. Osofsky [1] . Let Ω_α denote the first ordinal of cardinality $\aleph_\alpha$. Set $\Omega_{-1} = 1$ and consider -1 to be less than every ordinal.

<u>2·3. THEOREM</u> (OSOFSKY) Let R be a ring with unity. Let α be a non-zero ordinal and $M = \bigcup_{\beta<\alpha} y_\beta R$ be a unitary right R-module such that, for every $y \in M$ and $r \in R$, $yr = 0$ implies either $y = 0$ or $r = 0$. Further, let $y_\gamma R \subset y_\beta R$ if $\gamma < \beta < \alpha$. Then, for $n \geq -1$,

$$\text{pd } M = n + 1$$

if and only if the set of all ordinals $< \Omega_n$ is cofinal in the set $\{y_\beta R : \beta < \alpha\}$ ordered by inclusion.

<u>2·4. Proof of theorem 2·1.</u> The first two assertions are immediate consequences of 2·2, 2·3 and the global dimension theorem. See Jans [1 , page 56] .

Now suppose that $A = Rx_0$ is a proper prime ideal of R such that $A^{[\omega^1]} = \bigcap_{n \in Z^+} Rx_0^n \neq (0)$. Let $A^{[\omega^1]} = Rx_1$. As shown in the proof of lemma 2·2, there exists a unit u in R such that

$x_1 = ux_1 x_0$ so that $x_1 = u^n x_1 x_0^n$ for every $n \in Z^+$. Consider the sequence $\{u^n x_1 R : n \in Z^+\}$ of principal right ideals of R . Since $u^{n+1} x_1 x_0 = u^n x_1$, we have

$$u^n x_1 R \subseteq u^{n+1} x_1 R \ , \ n \in Z^+ \ .$$

If $u^n x_1 R = u^{n+1} x_1 R$ then there exists $r \in R$ such that $u^{n+1} x_1 = u^n x_1 r$ i.e., $x_1 = u^{-1} x_1 r = x_1 x_0 r$, which yields $1 = x_0 r$. Since R is a domain, x_0 is a unit, so $A = R$, contrary to our hypothesis. By 2·3, the homological dimension of $\bigcup_{n \in Z^+} u^n x_1 R$ is 1 ; so, by the global dimension theorem, r.gl. dim $R \geq 2$. This completes the proof.

References.

Theorem 2·1 is new; it is applicable to domains constructed in Jategaonkar [4] .

NOTES

Firstly, we state some open problems.

1. Is every simple pli-ring a pri-ring?

2. Is every primary pli-ring with a small prime radical an epimorphic image of a primary pli-ring with large prime radical?

3. Is every primary pli-ring with large prime radical an epimorphic image of a prime pli-ring?

4. Let R be a prime pli-ring. If $R \cong M_n(A)$ and $R \cong M_n(B)$ where A, B are domains, is it true that $A \cong B$?

5. Let R be a prime pli-ring and let Q be the simple Artinian l.q. ring of R. If there exists a monomorphism ρ: $Q \to R$, it is true that $R \cong M_n(A)$ where A is a pli-domain?

6. Let A be a domain such that $M_n(A)$ is a prime pli-ring. Is A a ipli-ring?

7. Is every ideal in a prime pli-ring a product of prime ideals?

8. Does every primary pli-ring (with large prime radical) have a l.q. ring?

If the reader has read the monograph before reading these notes, he will have noticed that we have not given any illustrative examples. Here are a few references: to see that most of the complications in chapters II and III are necessary, take a glance at Jategaonkar [3,4]. To get some feeling for twisted polynomial rings, consult examples in Jacobson [3, chapter 3] and examples in Divinski [1]. For an example of a nasty pli-domain, see Cohn [3]. For some reasonable left Noetherian ipli-rings, see Clark [1]. To get convinced that general left Noetherian rings are appalling, consult Small [4].

REFERENCES

K. Asano

[1] 'Quotient bildung und schiefringe'. J. Math. Soc. Japan, 1 (1949), 73-78.

H. Bass

[1] 'K-theory and stable algebra'. Institut des Hautes Scientifiques; Publ. Math. No. 22, (1964).

[2] 'Algebraic K-theory'. W. A. Benjamin, Inc., New York. 1968.

R. A. Beauregard

[1] 'Infinite primes and unique factorization in a principal right ideal domain'. Trans. Amer. Math. Soc., (to appear).

E. Cartan and S. Eilenberg

[1] 'Homological Algebra'. Princeton Univ. Press, Princeton, New Jersey. 1956.

S. U. Chase

[1] 'Direct products of modules'. Trans. Amer. Math. Soc. 97 (1960), 457-473.

W. E. Clark

[1] 'Murase's quasi-matrix rings and generalizations'. Sci. paper of the College of Education, Univ. Tokyo, 18 (1968), 99-109.

P. M. Cohn

[1] 'On a generalization of the Euclidean algorithm'. Proc. Cambridge Phil. Soc., 57 (1961), 18-30.

[2] 'On the embedding of rings in skew fields'. Proc. London Math. Soc., (3) 11 (1961), 511-530.

[3] 'Quadratic extension of skew fields'. Proc. London Math. Soc., (3) 11 (1961), 531-556.

[4] 'A remark on matrix rings over free ideal rings'. Proc. Cambridge Phil. Soc., 62 (1966), 1-4.

J. H. Cozzens

[1] 'Homological properties of the ring of differential polynomial'. (to appear).

C. W. Curtis

[1] 'A note on non-commutative polynomials'. Proc. Amer. Math. Soc., 3 (1952), 965-969.

N. J. Divinsky

[1] 'Rings and radicals'. Math. expositions No. 14, University of Toronto Press, 1965.

C. Faith

[1] 'Lectures on injective modules and quotient rings'. Lecture notes in Math.; Springer-Verlag, Berlin, Heidelberg, New York.

C. Faith and Y. Utumi

[1] 'On Noetherian prime rings'. Trans. Amer. Math. Soc. 114 (1965), 53-60.

A. W. Goldie

[1] 'The structure of prime rings under ascending chain condition'. Proc. London Math. Soc., 8 (1958), 589-608.

[2] 'Semi-prime rings with maximum condition'. Proc. London Math. Soc., 10 (1960), 201-220.

[3] 'Non-commutative principal ideal rings'. Arch. Math., 13 (1962), 214-221.

[4] 'Rings with maximum condition'. Lecture notes, Yale University Press. 1961.

[5] 'Some aspects of ring theory'. Bull. London Math. Soc. 1 (1969), 129-154.

R. N. Gupta

[1] 'Characterizations of rings whose classical quotient rings are perfect rings'.

[2] 'Self-injective quotient rings and injective quotient modules'. Osoka J. Math. 5 (1968), 69-87.

R. N. Gupta and F. Saha

[1] 'A remark on a paper of Small'. J. Math. Sci., 2 (1967), 7-16.

R. Hart

[1] 'Simple rings with uniform right ideals'. J. London Math. Soc., 42 (1967), 614-617.

I. N. Herstein

[1] 'Topics in ring theory'. Math. lecture Notes; Univ. Chicago Press. 1965.

I. N. Herstein and L. Small

[1] 'Nil rings satisfying certain chain conditions'. Canad. J. Math., 16 (1964), 771-776.

[2] 'Nil rings satisfying certain chain conditions: An addendum'. Canad. J. Math. 18 (1966), 300-303.

N. Jacobson

[1] 'A note on non-commutative polynomials'. Ann. of Math. (2) 35 (1934), 209-210.

[2] 'Theory of rings'. Amer. Math. Soc., Math Surveys, vol. 1, Providence, R. I., 1943.

[3] 'Structure of rings'. Amer. Math. Soc. Coll. publ., vol. 37 (revised edition), Providence, R. I., 1964.

J. P. Jans

[1] 'Rings and homology'. Holt, Rinehart and Winston, New York. 1964.

[2] 'On orders in Quasi-Frobenius rings'. J. Algebra, 7 (1967), 35-43.

A. V. Jategaonkar

[1] 'Left principal ideal rings'. Thesis (unpublished). Univ. of Rochester, Rochester, N. Y.; 1968.

[2] 'An example in principal left ideal rings' (unpublished).

[3] 'Left principal ideal domains'. J. Algebra, 8 (1968), 148-155.

[4] 'A counter-example in ring theory and homological algebra'. J. Algebra 12 (1969), 418-440.

[5] 'Structure of left principal ideal rings'. Notices Amer. Math. Soc. 15 (1968), 217.

[6] 'Local left Noetherian ipli-rings'. Bull. Amer. Math. Soc., 75 (1969), 514-516.

[7] 'Rings with transfinite left division algorithm'. Bull. Amer. Math. Soc., 75 (1969), 559-561.

[8] 'Orders in Artinian rings'. (To appear).

[9] 'Non-isomorphic Noetherian rings with isomorphic matrix rings'. (To appear).

[10] 'Ore domains and free algebras'. Bull. London Math. Soc., 1 (1969), 45-46.

R. E. Johnson

[1] 'Principal right ideal rings'. Canad. J. Math. 15 (1963), 297-301.

[2] 'Unique factorization in principal right ideal domains'. Proc. Amer. Math. Soc., 16 (1965), 526-528.

J. Lambek

[1] 'Lectures on rings and modules'. Blaisdell Publ. Co. Waltham, Mass., 1966.

C. Lanski

[1] 'Nil subrings of Goldie rings are nilpotent'. (To appear)

L. Levy

[1] 'Torsion-free and divisible modules over non-integral domains'. Canad. J. Math., 15 (1963), 132-151.

J. Levitski

[1] 'On nil subrings'. Israel J. Math., 1 (1963), 215-216.

L. Lesieur and R. Croisot

[1] 'Sur les anneaux premiers Noetheriens a gauche', Ann. Sci. Ec. Norm. sup. 76 (1959), 161-183.

N. H. McCoy

[1] 'The Theory of rings'. MacMillan, New York. 1964.

A. C. Mewborn and C. N. Winton

[1] 'Orders in self-injective semi-perfect rings'. J. Algebra 13 (1969), 5-9.

G. Mitchler

[1] 'On maximal nilpotent subrings of right Noetherian rings'. Glasgow Math. J., 8 (1967), 89-101.

O. Ore

[1] 'Theory of non-commutative polynomials'. Ann. of Math., 34 (1933), 480-508.

B. L. Osofsky

[1] 'Global dimension of valuation rings'. Trans. Amer. Math. Soc., 127 (1967), 136-149.

C. Procesi and L. Small

[1] 'On a theorem of Goldie'. J. Algebra, 2 (1965), 80-84.

J. C. Robson

[1] 'Artinian quotient rings'. Proc. London Math. Soc., (3) 17 (1967), 600-616.

[2] 'Rings in which finitely generated right ideals are principal'. Proc. London Math. Soc., (3) 17 (1967), 617-628.

[3] 'Pri-rings and ipri-rings'. Quart. J. Math. Oxford (2) 18 (1967), 125-145.

[4] 'Non-commutative Dedekind rings'. J. Algebra, 9 (1968), 249-265.

W. Sierpinski

[1] 'Cardinal and ordinal numbers'. Polska Acad. Nauk, Monografie Math., vol. 34. Warszawa, 1958.

L. W. Small

[1] 'Orders in Artinian rings'. J. Algebra, 4 (1966), 13-41.

[2] 'Orders in Artinian rings: Corrections and addendum'. J. Algebra, 4 (1966), 505-507.

[3] 'Orders in Artinian rings, II'. J. Algebra, 9 (1968), 266-273.

[4] 'On some questions in Noetherian rings'. Bull. Amer. Math. Soc. 72 (1966), 853-857.

R. C. Shock

[1] to appear.

[2] 'Nil ideals of rings satisfying maximum condition on right annihilators'. Notices AMS, 16 (1969), 806.

D. A. Smith

[1] 'On semi-groups, semi-rings and rings of quotients'. J. Sci. Hiroshima Univ. Ser A-I, 30 (1966), 123-130.

T. D. Talintyre

[1] 'Quotient rings of rings with maximum condition for right ideals'. J. London Math. Soc. 38 (1963), 439-450.

[2] 'Quotient rings with minimum condition on right ideals'. J. London Math. Soc. 41 (1966), 141-144.

Y. Utumi

[1] 'A theorem of Levitski'. Amer. Math. Monthly, 70 (1963), 286.

K. G. Wolfson

[1] 'Isomorphisms of the endomorphism ring of a free module over a principal left ideal domain'. Michigan Math. J., 9 (1962), 69-75.

O. Zariski and P. Samuel

[1] 'Commutative algebra'. vol. 1. Von Nostrand Co., Inc., New York; 1958.

INDEX

Offsetdruck: Julius Beltz, Weinheim/Bergstr